STEVE
DICKMAN

Geophysical Monograph Series

Including

IUGG Volumes
Maurice Ewing Volumes
Mineral Physics Volumes

GEOPHYSICAL MONOGRAPH SERIES

Geophysical Monograph Volumes

1 **Antarctica in the International Geophysical Year** *A. P. Crary, L. M. Gould, E. O. Hulburt, Hugh Odishaw, and Waldo E. Smith (Eds.)*
2 **Geophysics and the IGY** *Hugh Odishaw and Stanley Ruttenberg (Eds.)*
3 **Atmospheric Chemistry of Chlorine and Sulfur Compounds** *James P. Lodge, Jr. (Ed.)*
4 **Contemporary Geodesy** *Charles A. Whitten and Kenneth H. Drummond (Eds.)*
5 **Physics of Precipitation** *Helmut Weickmann (Ed.)*
6 **The Crust of the Pacific Basin** *Gordon A. Macdonald and Hisashi Kuno (Eds.)*
7 **Antarctica Research: The Matthew Fontaine Maury Memorial Symposium** *H. Wexler, M. J. Rubin, and J. E. Caskey, Jr. (Eds.)*
8 **Terrestrial Heat Flow** *William H. K. Lee (Ed.)*
9 **Gravity Anomalies: Unsurveyed Areas** *Hyman Orlin (Ed.)*
10 **The Earth Beneath the Continents: A Volume of Geophysical Studies in Honor of Merle A. Tuve** *John S. Steinhart and T. Jefferson Smith (Eds.)*
11 **Isotope Techniques in the Hydrologic Cycle** *Glenn E. Stout (Ed.)*
12 **The Crust and Upper Mantle of the Pacific Area** *Leon Knopoff, Charles L. Drake, and Pembroke J. Hart (Eds.)*
13 **The Earth's Crust and Upper Mantle** *Pembroke J. Hart (Ed.)*
14 **The Structure and Physical Properties of the Earth's Crust** *John G. Heacock (Ed.)*
15 **The Use of Artificial Satellites for Geodesy** *Soren W. Henricksen, Armando Mancini, and Bernard H. Chovitz (Eds.)*
16 **Flow and Fracture of Rocks** *H. C. Heard, I. Y. Borg, N. L. Carter, and C. B. Raleigh (Eds.)*
17 **Man-Made Lakes: Their Problems and Environmental Effects** *William C. Ackermann, Gilbert F. White, and E. B. Worthington (Eds.)*
18 **The Upper Atmosphere in Motion: A Selection of Papers With Annotation** *C. O. Hines and Colleagues*
19 **The Geophysics of the Pacific Ocean Basin and Its Margin: A Volume in Honor of George P. Woollard** *George H. Sutton, Murli H. Manghnani, and Ralph Moberly (Eds.)*
20 **The Earth's Crust: Its Nature and Physical Properties** *John G. Heacock (Ed.)*
21 **Quantitative Modeling of Magnetospheric Processes** *W. P. Olson (Ed.)*
22 **Derivation, Meaning, and Use of Geomagnetic Indices** *P. N. Mayaud*
23 **The Tectonic and Geologic Evolution of Southeast Asian Seas and Islands** *Dennis E. Hayes (Ed.)*
24 **Mechanical Behavior of Crustal Rocks: The Handin Volume** *N. L. Carter, M. Friedman, J. M. Logan, and D. W. Stearns (Eds.)*
25 **Physics of Auroral Arc Formation** *S.-I. Akasofu and J. R. Kan (Eds.)*
26 **Heterogeneous Atmospheric Chemistry** *David R. Schryer (Ed.)*
27 **The Tectonic and Geologic Evolution of Southeast Asian Seas and Islands: Part 2** *Dennis E. Hayes (Ed.)*
28 **Magnetospheric Currents** *Thomas A. Potemra (Ed.)*
29 **Climate Processes and Climate Sensitivity (Maurice Ewing Volume 5)** *James E. Hansen and Taro Takahashi (Eds.)*
30 **Magnetic Reconnection in Space and Laboratory Plasmas** *Edward W. Hones, Jr. (Ed.)*
31 **Point Defects in Minerals (Mineral Physics Volume 1)** *Robert N. Schock (Ed.)*
32 **The Carbon Cycle and Atmospheric CO_2: Natural Variations Archean to Present** *E. T. Sundquist and W. S. Broecker (Eds.)*
33 **Greenland Ice Core: Geophysics, Geochemistry, and the Environment** *C. C. Langway, Jr., H. Oeschger, and W. Dansgaard (Eds.)*
34 **Collisionless Shocks in the Heliosphere: A Tutorial Review** *Robert G. Stone and Bruce T. Tsurutani (Eds.)*
35 **Collisionless Shocks in the Heliosphere: Reviews of Current Research** *Bruce T. Tsurutani and Robert G. Stone (Eds.)*
36 **Mineral and Rock Deformation: Laboratory Studies—The Paterson Volume** *B. E. Hobbs and H. C. Heard (Eds.)*
37 **Earthquake Source Mechanics (Maurice Ewing Volume 6)** *Shamita Das, John Boatwright, and Christopher H. Scholz (Eds.)*
38 **Ion Acceleration in the Magnetosphere and Ionosphere** *Tom Chang (Ed.)*
39 **High Pressure Research in Mineral Physics (Mineral Physics Volume 2)** *Murli H. Manghnani and Yasuhiko Syono (Eds.)*
40 **Gondwana Six: Structure, Tectonics, and Geophysics** *Gary D. McKenzie (Ed.)*

41 Gondwana Six: Stratigraphy, Sedimentology, and Paleontoloty *Garry D. McKenzie (Ed.)*

42 Flow and Transport Through Unsaturated Fractured Rock *Daniel D. Evans and Thomas J. Nicholson (Eds.)*

43 Seamounts, Islands, and Atolls *Barbara H. Keating, Patricia Fryer, Rodey Batiza, and George W. Boehlert (Eds.)*

44 Modeling Magnetospheric Plasma *T. E. Moore, J. H. Waite, Jr. (Eds.)*

45 Perovskite: A Structure of Great Interest to Geophysics and Materials Science *Alexandra Navrotsky and Donald J. Weidner (Eds.)*

Maurice Ewing Volumes

1 Island Arcs, Deep Sea Trenches, and Back-Arc Basins *Manik Talwani and Walter C. Pitman III (Eds.)*

2 Deep Drilling Results in the Atlantic Ocean: Ocean Crust *Manik Talwani, Christopher G. Harrison, and Dennis E. Hayes (Eds.)*

3 Deep Drilling Results in the Atlantic Ocean: Continental Margins and Paleoenvironment *Manik Talwani, William Hay, and William B. F. Ryan (Eds.)*

4 Earthquake Prediction—An International Review *David W. Simpson and Paul G. Richards (Eds.)*

5 Climate Processes and Climate Sensitivity *James E. Hansen and Taro Takahashi (Eds.)*

6 Earthquake Source Mechanics *Shamita Das, John Boatwright, and Christopher H. Scholz (Eds.)*

Mineral Physics Volumes

1 Point Defects in Minerals *Robert N. Schock (Ed.)*

2 High Pressure Research in Mineral Physics *Murli H. Manghnani and Yasuhiko Syono (Eds.)*

Geophysical Monograph 46
IUGG Volume 1

Structure and Dynamics of Earth's Deep Interior

D. E. Smylie
Raymond Hide
Editors

American Geophysical Union
International Union of Geodesy and Geophysics

Geophysical Monograph/IUGG Series

Library of Congress Cataloging-in-Publication Data

Structure and dynamics of earth's deep interior.

(Geophysical monograph; 46/IUGG series ; 1)
"All Union Symposium U2 on 'Instability within the Earth and Core Dynamics' was held on Thursday, August 20 and Friday August 21, 1987 in Vancouver, Canada"—Preface
1. Earth—Core—Congresses. 2. Geodynamics—Congresses. I. Smylie, D. E. II. Hide, R. (Raymond), 1929– . III. All Union Symposium U2 on 'Instability within the Earth and Core Dynamics' (1987 : Vancouver, B.C.). IV. International Union of Geodesy and Geophysics. V. American Geophysical Union. VI. Series. QE509.I56 1987 551.1'12. 88-35014
ISBN 0-87590-450-5

Printed in the United States of America.

CONTENTS

PREFACE

All Union Symposium U2 on 'Instability within the Earth and Core Dynamics' was held on Thursday, August 20 and Friday, August 21, 1987 in Vancouver, Canada as part of the XIX General Assembly of the International Union of Geodesy and Geophysics. The Symposium was convened and organized by Raymond Hide, David Loper and D. E. Smylie. It involved the presentation of thirty papers in sessions entitled 'The Discovery of Core Modes:Observations and Theory', 'Core Structure and Flow', 'The International Program on the Study of the Earth's Deep Interior (SEDI)', 'Core Motions and Geomagnetism' and 'Hydromagnetics and Dynamo Theory'.

Much scientific excitement was generated by the lead-off invited paper presented by Professor Melchior on the superconducting gravimeter observations at Brussels. Subsequent papers dealt with the analysis of this data set and attempts to predict theoretically the expected response of the fluid core in realistic Earth models with the hope that a new field of long period core spectroscopy might emerge. There was a general consensus that an international effort is needed to provide a network of similar instruments distributed around the globe with agreed exchanges of observations and data reduction methods. Additional installations are either in place or planned in Canada, China, France, Germany and Japan.

Another highlight of the Symposium was the discussion of plans for SEDI, an international study of the Earth's Deep Interior which was adopted at the Vancouver General Assembly. As well as refocussing the efforts of dynamo theorists on using large scale vector computers to make progress on one of the most fundamental problems of geophysics, SEDI hopes to bring an international interest to bear on the problem of the Earth's heat engine as represented by the two great convective systems in the mantle and the fluid core. Thus SEDI will examine such topics as core-mantle interactions, structure and dynamics of the lower mantle including the D" layer, mantle plumes and the structure and evolution of the inner and outer cores, as well as the hydromagnetics and oscillations of the fluid outer core.

Continued progress in mapping the magnetic field and flows at the core-mantle boundary was reported in papers by Bloxham and Benton, and in dynamo theory in papers by Busse(invited), Szeto and Vishik. Lower mantle and core-mantle boundary structural results based on seismic methods were reported by Okal and Haddon, while papers on plume formation and the effect of mantle convection on lower boundary topography were presented by Olson and Baumgardner. The areas of core-mantle coupling, magnetic field variations and core motions were represented in papers by Hide(invited), Kakuta, Merriam, Nicolaysen, Okubo, Runcorn(invited) and Yukutake.

The following eighteen papers form the written record of the Symposium but we see its greater importance as marking a major rebirth of interest in the deep interior workings of the planet we call home. Through the new international project SEDI this renewed interest will be focussed and nurtured over the next several years.

As Editors, we have seen our role as facilitating and encouraging as complete a record of the proceedings of the Symposium as possible and although all of the papers are of a high standard and represent the best achievements in a broad range of countries, we have taken the view that the ultimate responsibility for scientific accuracy and quality rests with the Authors themselves. They are all to be congratulated for the great effort they have put into the production of manuscripts for these Proceedings.

Raymond Hide D. E. Smylie

HAVE INERTIAL WAVES BEEN IDENTIFIED FROM THE EARTH'S CORE?

P. J. Melchior[1], D. J. Crossley[2], V. P. Dehant[3] and B. Ducarme[1]

Abstract. 1986 saw two significant developments in long-period gravimetry. The first was the publication by Melchior and Ducarme [1986; MD] of a spectrum calculated from vertical gravity measurements from the superconducting gravimeter in Brussels. The record was analysed following two large, deep earthquakes and showed several peaks in the period range 13 to 16 hr. [MD] speculated that the largest peak at 13.9 hr. could be due to either an internal gravity wave in the outer core, or translational motion of the inner core, excited by earthquakes. The second development came when Aldridge and Lumb [1987; AL] claimed that the [MD] spectrum in fact contained 8 peaks which were close to the known period of inertial waves in a rotating sphere filled with homogeneous inviscid fluid.

We first review some useful results from the superconducting gravimeter as related to the Earth's core and discuss observations of the free core nutation. We also present the data reduction methods used to produce a new spectrum of the complete 4-year superconducting gravimeter data set (from May 1982 to December 1986) which includes the previous [MD] data as a subset. We identify 10 peaks in the record which seem to be above the spectral noise level of 4 ngals from frequencies 0.058 to 0.077 per hr (periods 17.24 to 13.00 hr), and select 7 earthquakes in that time, with seismic moments above 10^{20} N.m. We find enhanced spectra following the two large Fijii earthquakes (May, 1985), very similar in style to the previous results [MD], whereas following the large Mexican earthquake (September, 1985) the spectrum is much more disturbed with large peaks which show evidence of drift with time. The spectral peaks are not at the same frequencies from one earthquake to the next.

We attempt to interpret the new spectrum in terms of the inertial wave model. Evidence to date suggests that the largest computational error in the theoretical inertial-wave eigenperiods used by [AL] arises from neglect of the inner core. Laboratory observations and theoretical eigenperiods published by Aldridge [1972] for a spherical shell show considerable uncertainty about the sense and magnitude of the difference from the (exact) calculations for a full sphere and have to be treated as unreliable. More recent computations of eigenperiods of internal gravity waves in a strongly- stable core (where convergence is considerably more certain than the weak or neutral-stratification case) can be used to infer the effect of buoyancy and an inner core on the eigenperiods obtained from Poincaré's theory.

Comparison of the 10 peaks in the 4-year spectrum with the calculated periods of inertial waves with simple spatial structure in a full sphere do not strike us a showing clear evidence of correlation. We discuss the extent to which the agreement claimed by [AL] may still exist (e.g. the gravimeter time series is highly non-stationary which may negate the spectral analysis). We also present the case that the gravimeter results can be otherwise explained by ocean tidal resonances in the North sea or, alternatively, as the long-sought-for translational motion of the Earth's inner core.

Resolution of the mechanism causing the spectral peaks we observe clearly requires a global network of high quality measurements using superconducting gravimeters.

[1]Observatoire Royal de Belgique, Avenue Circulaire 3, B-1180 Bruxelles, Belgium.
[2]Geophysics Laboratory, McGill University, 3450 University Street, Montreal, Canada H3A 2A7.
[3]Institut d'Astronomie et de Géophysique G. Lemaître, Université Catholique de Louvain, 2, Chemin du Cyclotron, B-1348 Louvain-La-Neuve, Belgium.

The Core Mantle Boundary

Long series of tidal registrations with classical spring gravimeters and quartz tiltmeters of high quality, performed during the past twenty years or more at several permanent earth tide stations, have demonstrated without doubt the existence of a hydrodynamic resonance at a nearly diurnal frequency in the liquid core of the Earth [Melchior, 1966]. The frequency of this resonance can be theoretically calculated using a perfectly ellipsoidal model

TABLE 1. Diurnal Tides and Associated Nutations.

	Frequency (°per hr.)	Period (hr.)	Associated Nutation	
			Cycle per year	Period (days)
K_1 (36233) lunar (16817) solar	15.041069	23.9345	0	secular (forced precession)
FCN - Earth Models				
Jeffreys/Vicente	15.0747	23.8811	0.8171	447
Molodensky I	15.0732651	23.8833	0.7840	466
Molodensky II	15.0736125	23.8828	0.7924	461
Sasao/Wahr(1066A)	15.073641		0.7932	460.5
FCN - Observed				
VLBI[1]	15.0757	23.8795	0.8431	433.2 ±2
VLBI[2]	15.07555	23.8797	0.8397	435.0 ±2
Earth Tides[3]	15.0757 ±0.0005	23.8795	0.8431	433 ±7
Forced Nutation				
ψ_1 (423)	15.082135	23.8693	0.9999	365.2657
ϕ_1 (756)	15.123206	23.8045	2.000	182.622

[1] Without ocean tide correction. [2] With ocean tide correction
[3] Two superconducting gravimeters : Brussels, Frankfurt.

of the core boundary having the Clairaut-Radau hydrostatic flattening, e.g. 1/393 with model 1066A [Gilbert and Dziewonski, 1981], equivalent to 9 km difference between equatorial and polar radii. In this type of model there is no boundary layer nor dissipation. The frequency was found to be either 15.0732651°per hr. (Model I) or 15.0736125°per hr. (Model II) by Molodensky [1961]. A Free Core Nutation (FCN), as seen from inertial space, is associated with this resonant frequency with a period of 466 days, i.e. 15.0732651°- 15.041069°(the sideral frequency) = 0.321961°per hr. or 0.7727064°per day.

Resonance may happen when the oscillations of the boundary have a period very close to the period of free oscillations of the fluid in its container. In practise, the viscosity in the core limits the strength of the resonance. This is the case indeed for several tidal waves, specially for the sideral lunisolar wave K_1 (frequency 15.041069°per hr.) and for the elliptic wave of the solar part of K_1 (called the ψ_1 wave; frequency 15.082135°per hr.). Unfortunately this wave is of very small amplitude so that the classical instruments could not determine it with a precision sufficient to make a choice between different core models or obtain information about the dissipation in the core-mantle boundary layer. This wave ψ_1 is associated with an annual nutation (more precisely a short period precession as $\Delta\theta = 0$) of the axes of inertia and rotation of the Earth (15.082135°- 15.041069°= 0.041066°per hr., period 365.2657 days).

Recent measurements made in the last four years with two superconducting gravimeters at Brussels and at Frankfurt, as well as precise determination of the associated annual nutation by VLBI observations, have exhibited a shift of the observed frequency with respect to the frequency computed with hydrostatic models. This corresponds to a 433 days period for the free core nutation (Table 1).

The discrepancy could be explained by approximating the actual core-mantle boundary by a non-hydrostatic flattening of some 500 meters larger than 9 km. This result is consistent with an interpretation of the low order coefficents in the Earth's gravity field in terms of deformation of this boundary with bumps of this size on it. It could also imply some change in the radial pressure at the core-mantle boundary.

The Q (quality factor, or damping) of this FCN eigenmode has also been calcuated from the superconducting gravimeters series. It was found to be 2305 ± 675 at Brussels and 3131 ± 826 at Frankfurt with as mean value 2767 ± 529 [Neuberg at al., 1987]. Also shown in Table 1 are the amplitudes, indicated in parentheses, for the K_1 and ψ_1 tidal waves.

Waves in the Core ?

The widely used spring gravimeters reach, in good conditions of installation, a precision slightly better than one microgal (10^{-9} of g). At this level no indication of any kind of gravity oscillations other than those present at strict tidal frequencies have ever been observed.

The possibility of observing internal oscillations of the core or inner core oscillations at the Earth's surface should

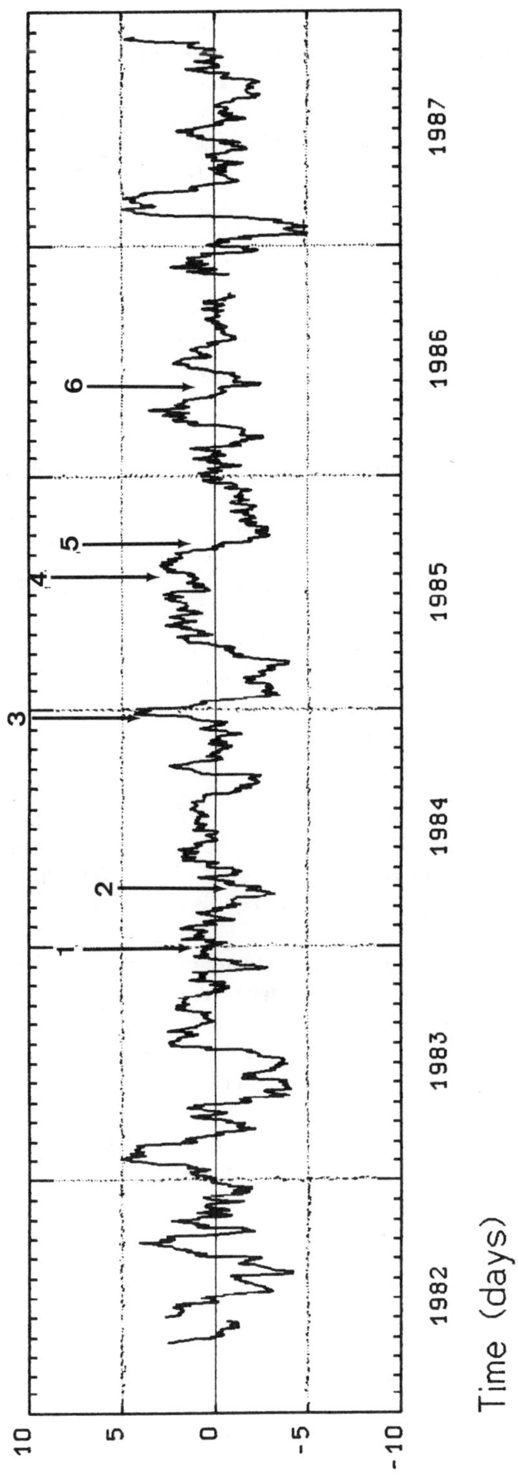

Fig. 1. 4 years of residuals of the superconducting gravimeter from Brussels. These data are obtained by subtracting from the original data the tidal signal, the instrumental drift, the effects of polar motion and an annual term. The numbers correspond to the events reproduced in Table 3.

therefore involve continuous gravity measurements at least at the nanogal level (10^{-12} of g). This precision became a possibility with the development and installation of superconducting gravimeters. A search was made in this direction by Melchior and Ducarme [1986; MD].

A superconducting gravimeter GWR (Goodkind, Warburton and Reineman) was installed at the Royal Observatory of Belgium, Brussels (latitude =50°47'55" N, longitude = 4°21'29" E, elevation 104 m) in May 1981 and has been recording without interruption during more than four years (1596 days). The continuous record of gravity variations from this period is shown in Figure 1. The tidal analysis of this record has been made by the least squares method with the Venedikov filters and the Cartwright-Taylor-Edden (CTE) potential which allowed the determination of the response (amplitude factor and phase) for not less than 20 different tidal frequencies in the diurnal band, 13 in the semi-diurnal band, 3 in the ter-diurnal band and 5 long period waves.

The effect of the atmospheric pressure has been removed hour by hour by determining first the impulse response according to the De Meyer [1982] procedure which provides the correction formula at Brussels :

$$\Delta g = -0.26277P(t) - 0.07785P(t-1) - 0.01669P(t-2),$$

in μgal, where P is pressure in H millibar. The perturbation of the centrifugal force due to the effect of polar motion has been removed by applying a correction

$$\Delta g = \delta[\omega^2 r(x \cos\lambda - y \sin\lambda) sin 2\phi]$$

where (x, y) are the pole coordinates as given by the BIH, and δ was taken equal to 1.16.

A spectral analysis of the data, after removal of the atmospheric pressure and polar motion effects, has revealed the presence of a small peak of 11 ngal amplititude at 13.9 h. To check the reality of such a peak it is necessary to eliminate all tidal components. We have calculated 'residues' by subtracting, from the original data, a tidal signal calculated with the CTE potential completed to 830 terms according to Xi Qinwen [1986] and using the amplitude factors and phases obtained from the least squares analysis of the original data themselves.

A considerable decrease in the noise level results from the fact that we introduced the appropriate response (δ factor 1.06) for those diurnal and semi-diurnal components which derive from the third order term of the tidal potential. As a result the M_2 wave, which has an observed amplitude of $35\mu gal$, is reduced to an amplitude of $0.017\mu gal$ and the K_1 diurnal wave reduced from $49\mu gal$ to $0.023\mu gal$, that is about 5 x 10^{-4} of their amplitudes. The main tidal spectral lines are situated in the 11.97 - 12.91 hr. and 20.34 - 26.87 hr. bands but the Xi potential introduces very small contributions in wider bands : 11.15

[Handwritten margin note:] These are all subject to Revision !

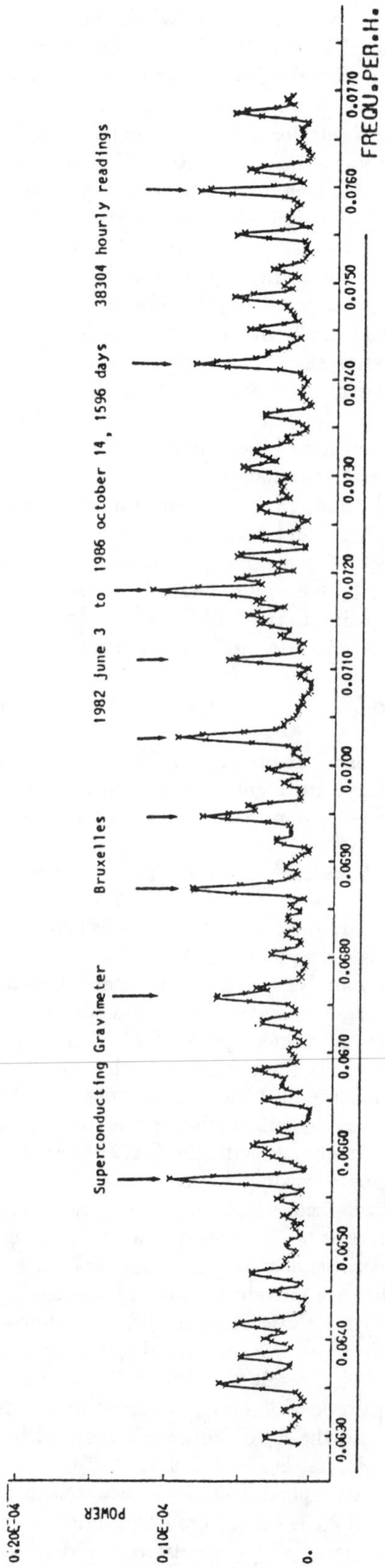

Fig. 2. Portion of spectrum of Fig. 1. The data were corrected for all known tidal components.

- 13.67 hr. and 20.25 - 31.88 hr. which have been eliminated by our procedure for calculation of the residues. A spectral analysis of these final residues shows that the background noise in the frequency band from 13 to 19 hr. is less than 4 ngal and that nine spectral lines merge from this noise (Figure 2, Table 2). The biggest of these peaks corresponds to a period of 13.924 hr.; the next largest at 15.225 hr.

TABLE 2. Superconducting Gravimeter in Brussels. Power spectrum of 1440 days observations from 1982 till 1986, no gaps. Small peaks observed between the tidal frequency bands

Frequency (cycle per hr.)	Period (hr.)	Amplitude x 10^{-3}	Power x 10^{-6}
0.05839	17.126	3.626	9.492
0.06568	15.225	3.653	9.676
0.06757	14.801	3.043	6.585
0.06869	14.558	3.390	8.157
0.06947	14.394	3.167	7.605
0.07028	1.118	3.161	9.281
0.07109	14.066	2.932	5.888
0.07182	13.924	4.031	11.061
0.07416	13.484	3.515	8.225
0.07595	13.166	3.627	8.003

The source of excitation of such oscillations is supposed to be due to seismic energy release [Mullan, 1973] while their viscous damping is considered to correspond to a spin-down time scale of about 1 yr. [Friedlander and Siegmann, 1982]. We suppose that strong deep earthquakes can deform the core-mantle interface and consequently tilt its axis of inertia with respect to its axis of rotation. The perturbation is the centrifugal force has a tesseral distribution, can excite a free nutation (FCN) and generates oscillations with a longitude wave number $m = 1$.

Since January 1982 there were only four deep earthquakes with a seismic moment higher than $10^{20} N.m$ (Table 3). If this is really the excitation source, one cannot expect that the oscillations in the core will remain in phase for a period of more than 1 yr. so that the spectral analysis of a very long interval may not allow the detection of such oscillations. One should rather analyse intervals of few months - five or six - immediately after deep strong earthquakes.

It does not seem indeed that shallow earthquakes have an influence : we performed a spectral analysis at 6 month intervals, starting immediately after the 7.7 magnitude earthquake of May 26, 1983, which occurred in Japan at a depth of 24 km, and did not observe any change of amplitude in the spectral lines. However, just after the two deep earthquakes of Hindu Kush (December 30, 1983)

TABLE 3. Large Moment Earthquakes 1982-86.

Event	Year	Month	Day	ϕ (lat.)	λ (long.)	Depth (km)	Magnitude	Seismic Moment (Newton metre)
1	1983	12	30	36.42 N	70.75 E	209	6.2	1.5×10^{20}
				(Hindu Kush)				
2	1984	3	6	9.5 N	138.92 E	454	6.1	1.4×10^{20}
				(South of Honshu)				
3	1984	11	20	5.15 N	125.12 E	202	6.4	2.1×10^{20}
				(Mindinao)				
4	1985	7	29	36.19 N	70.89 E	101	6.7	1.5×10^{20}
				(Hindu Kush)				
5	985	9	19	18.18 N	102.57 W	33	7.0	1.1×10^{21}
				(Mexico)				
6	1986	5	26	21.72 S	17.25 W	604	5.9	2.1×10^{19}
				20.07 S	178.72 E	553	6.8	5.6×10^{19}
				(South of Fijii)				

and Mindanao (November 20, 1984) there was an increase in the peak amplitudes which disappears after four or five months [Figure 3; MD]. The fault plane solutions of these two earthquakes are of the dip slip type. Results are given in the Table 4 where the effect of the windows used in the analysis is shown, the Parzen window giving a broadened peak. It is interesting to note that, analysing a 6 month period just before these earthquakes, we found no evidence of the 13.9 hr peak (see in Figure 3 the spectrum of 180 days starting on 1984.05.24).

Zürn et al. [1987] have recently presented a comment on these observations, using the registrations of a same instrument (GWR superconducting gravimeter) installed at Bad Homburg (Germany), a station distant from Brussels by 320 kilometers. The record of this instrument ended April 30, 1984 due to an instrumental failure so that only one of the deep earthquakes (Hindu Kush) could be examined from a 60 day record starting 60 days after the earthquake. It is thus simultaneous with the second half of our 120 day record, the spectrum of which is given in [MD]. Zürn et al. did not find any peak at a 13.9 hr. period.

We now observe a similar gravity effect following the two deep and almost simultaneous earthquakes of South

TABLE 4. Spectral Analysis after Two Earthquakes.

Event	Seismicity	Record Length (days)	Cosine Window A^1	Cosine Window T^2	Parzen Window A^1	Parzen Window T^2	Noise Background[1]
1	Hindu Kush	120	18.8	13.77	13.0	13.78	4.0
	1983-12-30				10.9	13.71	
	m=7.2	150	17.4	13.92	10.5	13.85	3.5
	Depth 222 km				12.0	13.79	
	(Fig.1)	180	17.3	13.91	9.4	13.87	3.0
					9.3	13.91	
		240	11.6	13.91	8.6	13.91	2.5
			11.6	13.94	6.5	13.88	
3	Mindanao	120	16.8	13.98	8.8	14.05	3.5
	1984-11-20				11.4	13.98	
	Depth 202 km	150	18.1	13.95	8.5	14.00	3.0
					10.2	13.95	
	(Fig.2)	180	14.9	13.91	8.8	13.93	3.0
			14.1	13.87	8.8	13.91	
		240	15.1	13.85	9.1	13.88	4.0
					8.8	13.85	

[1]Amplitudes A and noise are given in ngals. [2]Periods T are given in hr.

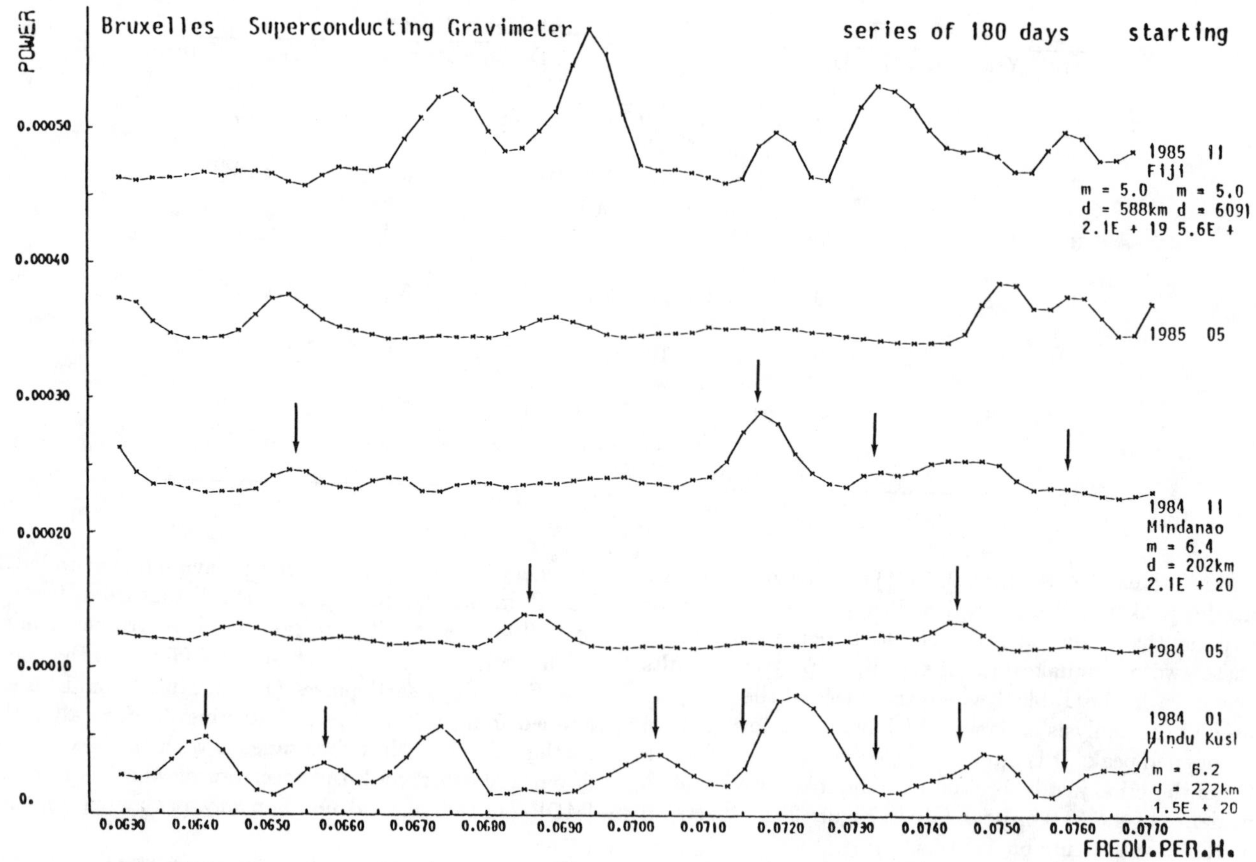

Fig. 3. Spectra before and after the Hindu Kush and Mindinao earthquakes, after [MD]. Note the disappearance of the 13.9 hr peak for spectra starting 6 months after the events. The small vertical arrows show the periods identified by [AL] as being inertial waves. [AL] do not state from which spectrum of [MD] their identifications were made, though it was learned later (Aldridge, personal communication) that they were taken from the spectrum beginning 60 days after the events, not the spectra shown here.

Fiji (26 May 1986) as shown also in Figure 4. We also made four series of spectral analyses covering each 180 days of residues of our tidal measurements with the superconducting gravimeter starting after the very strong Mexican earthquake (see Table 3). The four series (Figure 5) appear much more disturbed than the preceeding series of 1984 and 1985 (Figure 3). We also note that the peaks are suddenly enhanced when we shift the analysis by 30 days after the Mexican earthquake (series starting on 18 October 1985 - Figure 5). In March 1986, the spectrum returns to a relatively flat state (series starting on 19 March 1986 - Figure 6).

If such oscillations were to be ascribed to inner core translations (see below), as suggested a long time ago, a displacement of the inner core of only 3 cm would be indeed sufficient to generate a signal of 10 ngals at the surface of the Earth [Won and Kuo, 1973; Smith, 1976].

Possible Environmental Effects

Figure 7 shows the results of spectral analyses of oceanic tides as observed at Oostende harbour which is the nearest harbour to Brussels. There is no coincidence of the spectral lines with those observed with the superconducting gravimeter. Similar analyses have been made for the atmospheric pressure at Brussels. The result is a very noisy spectrum.

The Identification of Inertial Waves

We comment briefly on the contention by Aldridge and Lumb [1987; AL] that the gravity peaks are to be interpreted as inertial waves in the core. Without speculating further on how such waves might be excited (or produce a gravity signal at the Earth's surface), we review the

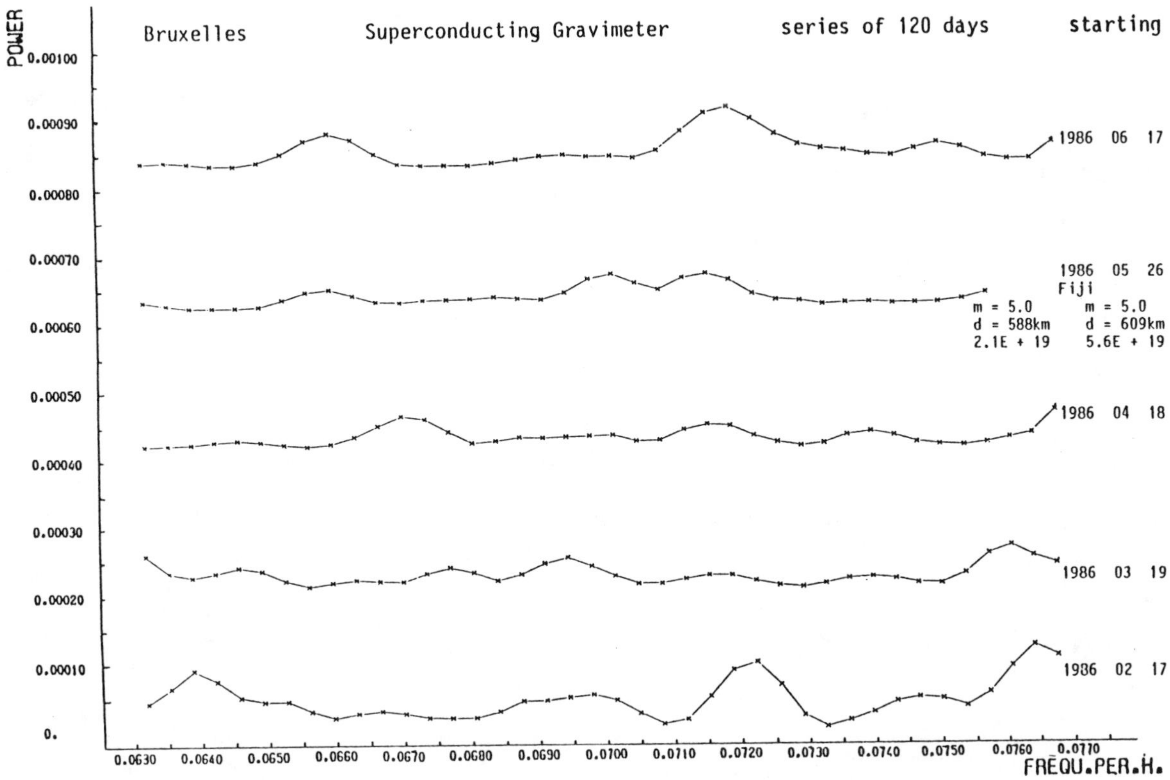

Fig. 4. Spectra of 4-month series at overlapping monthly intervals before, during and after the Fijii events of May, 1986.

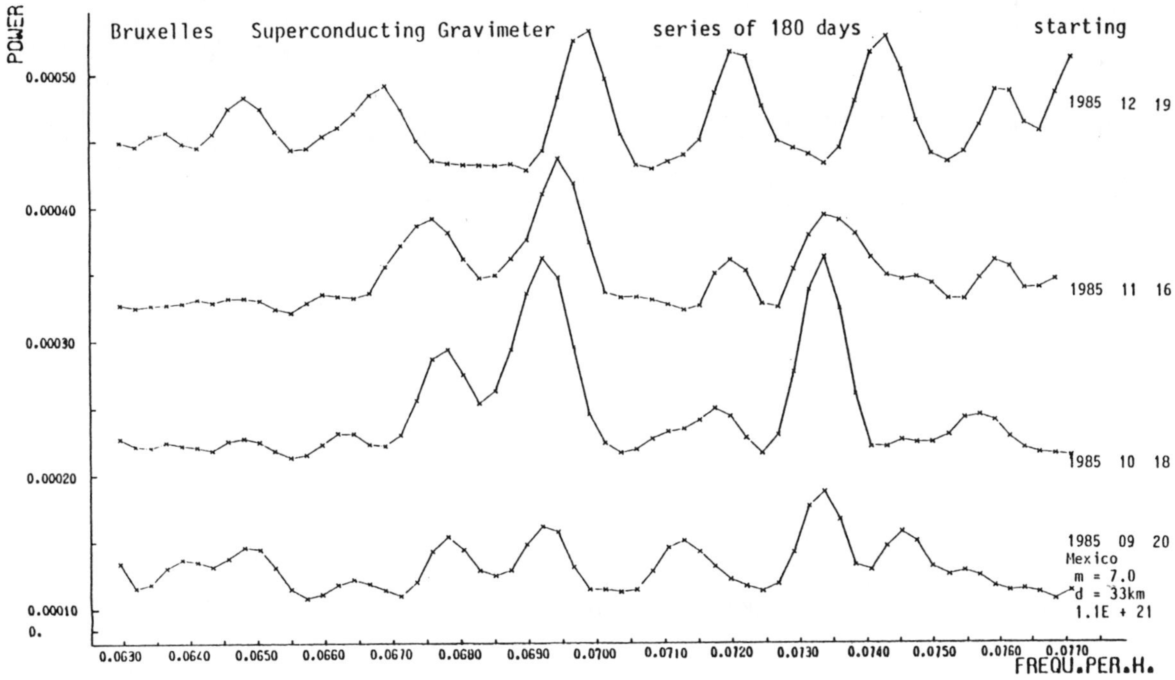

Fig. 5. Spectra of 6-month series at overlapping monthly intervals following the Mexican earthquake of September, 1985.

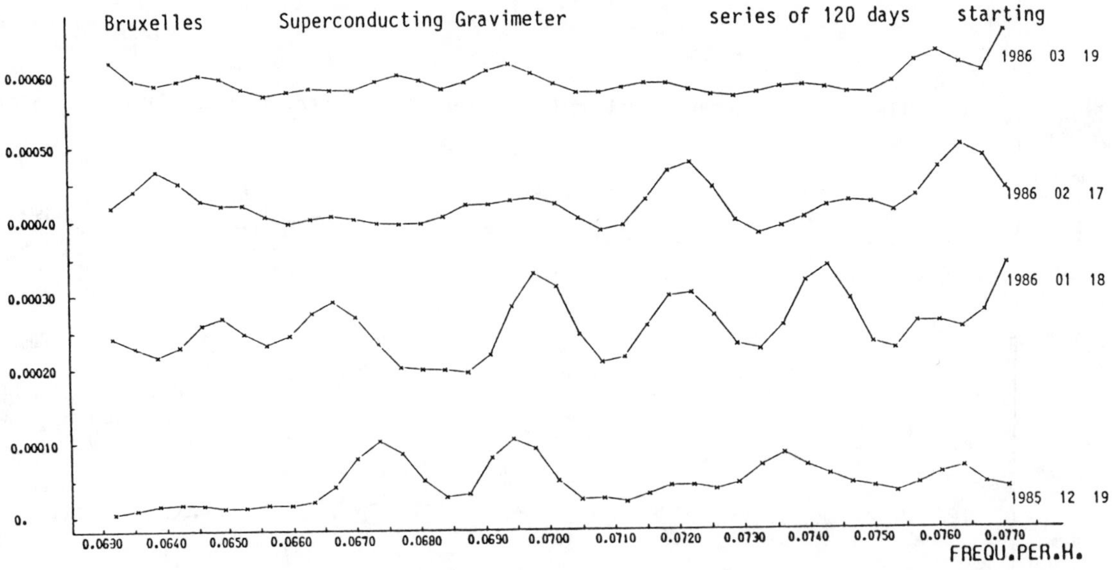

Fig. 6. Spectra of 4-month series at overlapping monthly intervals starting 3 months after the Mexican event.

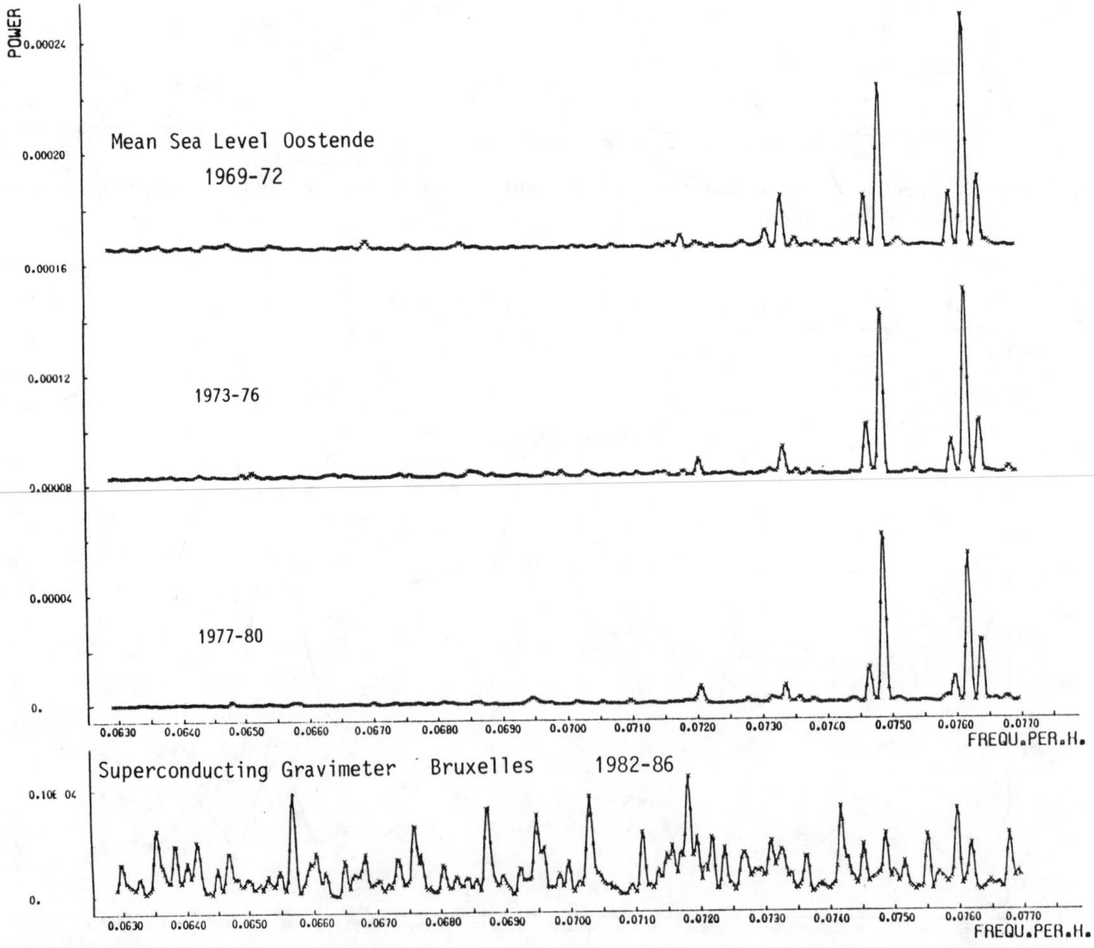

Fig. 7. Spectra of the mean sea level at Oostende for 3 time intervals showing a negligible sea level effect in the superconducting gravimeter data.

TABLE 5. Inertial Wave Periods.
Container is a fluid-filled spheroid of flattening 1/392.
• denotes waves selected by Aldridge and Lumb [1987].

frequency (per hr)	period (solar hr)	n	m	k		frequency (per hr)	period (solar hr)	n	m	k	
0.05608	17.831	6	2	3		0.06640	15.061	8	3	1	
0.05661	17.666	10	3	1		0.06765	14.781	10	3	4	
0.05666	17.648	8	2	0		0.06790	14.728	8	3	3	
0.05718	17.490	10	3	3		0.06854	14.591	10	4	-2	
0.05773	17.324	9	3	-3		0.06872	14.551	6	2	2	•
0.05814	17.200	10	3	-1		0.06877	14.540	8	3	-1	
0.05874	17.023	6	2	-1		0.06943	14.403	6	2	0	•
0.05876	17.019	8	2	4		0.06976	14.335	9	3	1	
0.06056	16.513	10	3	5		0.07111	14.063	9	3	3	
0.06089	16.422	8	3	-2		0.07141	14.003	4	2	1	•
0.06131	16.310	9	3	2		0.07166	13.954	9	4	-1	
0.06165	16.221	7	2	1		0.07223	13.845	10	4	1	
0.06180	16.180	9	3	0		0.07249	13.794	7	3	2	
0.06231	16.049	10	3	-3		0.07289	13.720	7	3	0	•
0.06260	15.975	5	2	2		0.07341	13.621	10	4	3	
0.06316	15.832	3	1	1	•	0.07378	13.553	10	4	-1	
0.06324	15.812	7	2	3		0.07498	13.336	8	3	2	
0.06396	15.636	9	3	4		0.07522	13.294	8	3	0	•
0.06400	15.626	5	2	0	•	0.07549	13.246	5	2	1	•
0.06471	15.454	7	3	-1		0.07672	13.035	9	4	2	
0.06529	15.317	10	3	2		0.07687	13.009	9	4	0	
0.06529	15.315	9	3	-2		0.07781	12.852	6	3	1	
0.06529	15.315	9	3	-2		0.07797	12.825	10	4	2	
0.06562	15.240	10	3	0							

theoretical spectrum of inertial waves computed for a rotating shell of fluid. The most general solution to date, for a spheroid of arbitrary ellipticity filled completely with fluid, was obtained by Kudlick [1966] with an allowance also for viscosity.

As indicated in [AL], the solution for the inviscid frequencies is obtained by finding the roots of a simple combination of Associated Legendre Functions of degree n and azimuthal order k. The m'th root (counted from the origin $x = \cos\theta = 0$) provides the radial quantum number for the wave. We have repeated the calculation for all combinations of (n,m,k) up to (10,5,5) where k is allowed to take on negative values to represent the westwards travelling waves. Equivalently one can find all roots in the range $-1.0 \leq x \leq 1.0$ with $k > 0$. The inertial wave periods agree with all previous authors [Kudlick, 1966; Greenspan, 1969; Aldridge and Toomre 1969; AL], with one exception. The mode (4,1,1) quoted by Greenspan (Table 2.1, p.66) has the eigenvalue for mode (4,3,1) and we find that mode (4,3,1) is a westwards wave denoted here by a negative azimuthal number, i.e. (4,1,-1). We remark also that the notation for the modes is sometimes confused, especially between Kudlick and Greenspan (these problems were also pointed out by M. Rochester [personal communication, 1987]).

For the sake of clarity, we quote in Table 5 all the inertial wave periods up to our maximum (n, m, k), between the frequencies 0.056 and 0.078 /hr, corresponding to the frequency range of the peaks identified in the gravimeter record. The frequencies have been computed for a core flattening of 1/392, that used by [AL], and also corrected to a solar rotation period (instead of 24 hr. which would give sidereal frequencies).

In order to assess the likelihood of these modes being excited, we use a simple scaling of the form $1/(n + m + \mid k \mid)$ to be an excitation amplitude. This empirical choice gives a greater weight to modes with a simple spatial structure, i.e. small values of (n, m, k), as such modes might be expected to be more easily excited [AL]. A plot of the theoretical spectrum appears in the bottom half of Figure 8, with the modes selected by [AL] denoted by symbols.

The new spectral analysis (Figure 2, Table 2) has been plotted in the upper part of Figure 8, with the distance from the top representing approximate spectral power (Table 2). On the basis that the spectral peaks corre-

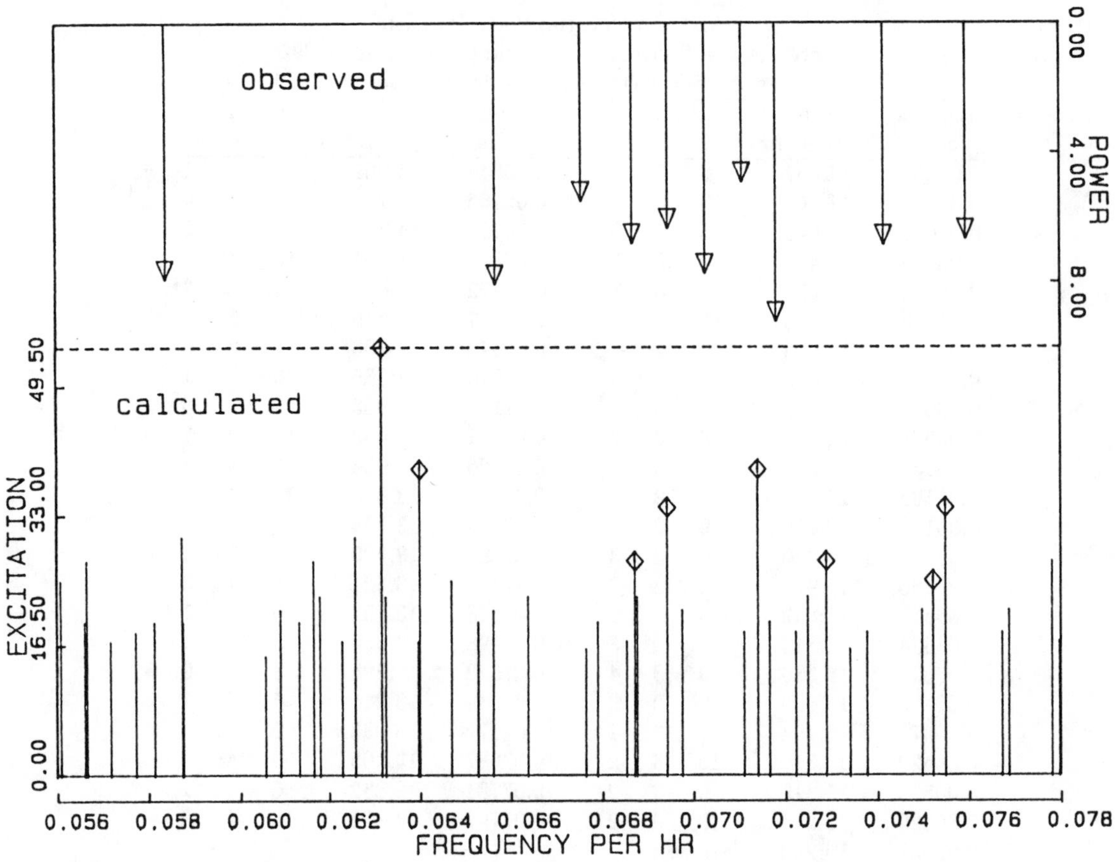

Fig. 8. Inertial wave periods compared. The calculations are repeated from the formula of Kudlick [1966] for a full sphere. The observations are the 10 prominent peaks in the composite 4-year record at Brussels (Table 2).

spond to easily excited inertial waves [AL], the two portions of Figure 8 should agree.

We note that to first order, the two data sets do not match. One possible explanation why they do not is that the calculated eigenfrequencies have to be modifed for the effect of the inner core. Though Aldridge [1972] studied the effect of the inner sphere, his results are ambiguous with respect to the sense and magnitude of the correction to the full sphere eigenfrequencies. On the other hand, calculations for internal gravity waves and Rossby waves in a thin shell [Longuet-Higgins, 1968] show that the eigenfrequencies become smaller as the shell thickness increases, though whether this effect continues down to zero radius for the inner sphere is not known.

By the above reasoning we may therefore imagine displacing (in Figure 8) the lower portion (calculated frequencies) with respect to the upper (observed frequencies), to simulate the inner core correction (assuming reasonably that it will be of a consistent sense and size within this relatively confined frequency band). The reader can satisfy himself that such displacement does not significantly improve the top-bottom match. Of course if one relaxes the assumption that the largest amplitude gravity peaks should coincide with the inertial waves of lowest quantum numbers, there are certainly sufficient inertial modes to fit the gravity spectrum. If the upper limit to (n,m,k) is increased indefinitely, the inertial wave spectrum becomes in fact infinitely dense [Aldridge and Toomre, 1969; Crossley, 1984].

The remaining effect to be considered is whether it is reasonable to assume the whole gravity spectrum is stationary or whether the inertial wave periods are modified by other dynamical effects as suggested by [AL] On the basis of a simple comparison, however, it is by no means clear to us that the peaks are inertial waves. We are evidently still at a point where observation and theory both need further development to establish whether the connection between the core and the gravity record is real.

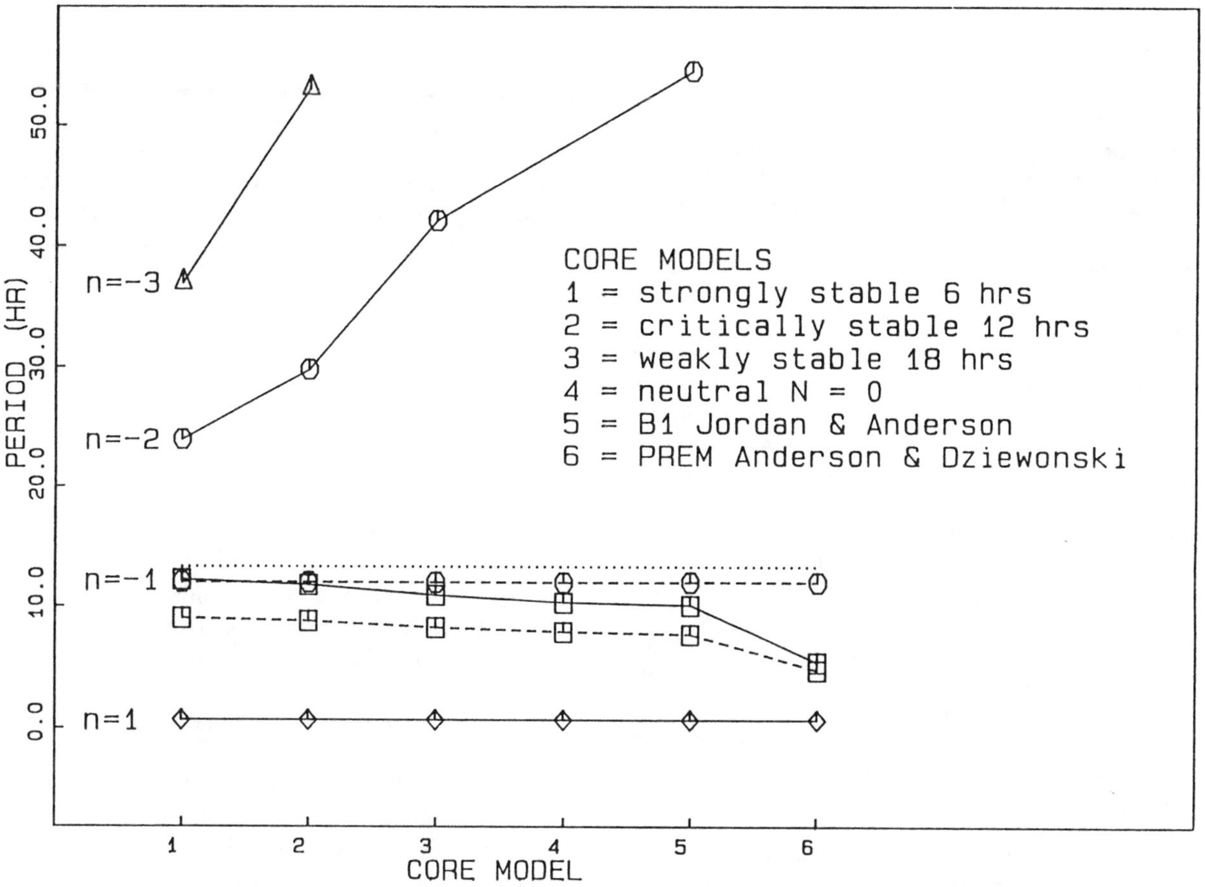

Fig. 9. Inner core translational periods for a variety of core models, both with (dashed lines; indicating inclusion of the next torsional term only) and without (solid lines) rotation. The horizontal dotted line is the projected limit of the periods for this level of truncation in a rotating Earth model for azimuthal number $m = 1$.

The Possibility of Inner Core Motion

As indicated earlier, the dominance of the 13.9 hr period in the previous studies [MD], suggests a mechanism with a single degree of freedom rather than the inertial wave spectrum. We therefore have repeated calculations similar to those of Crossley [1975] and Smith [1976] to establish whether inner core modes in the period range of interest are possible.

Inner core translation in a non-rotating Earth model is a spheroidal deformation denoted ${}_nS_1^m$, where here we follow the elastic normal mode identification scheme and denote (radial number,degree,order) by (n, l, m); for $l = 1, m = 0, \pm 1$. We have calculated the eigenfrequencies of several modes for $m = 1$ (the axial modes) for a suite of different core models, listed 1-6 in Figure 9. The designation $n = 1$ indicates the overtone and $n = -1, -2, \ldots$ the undertones [Crossley, 1975]. Solid lines are the periods without rotation, dashed lines the periods with the addition of the mode ${}_nT_2^m$, showing the drastic effect of Coriolis coupling on the eigenperiods. Shown are all modes found at periods less than 60 hr., with the possible omission of modes between the dotted line and the dashed line for $n = -2$ with rotation (shown as circles).

Without rotation, modes can have almost any periods between about 5hr and 60 hr depending on the Earth model. The effect of core stratification is seen to act differently for the $n = -1$ as opposed to the $n = -2$ and $n = -3$ modes. Perhaps the reason for this lies in the fact that the $n = -2, -3$ modes have more than one nodal point radially in the outer core [see Crossley, 1975, Figure 4] and are thus more influenced by core stratification. As can be seen by looking at the realistic Earth models B1 [Jordan and Anderson, 1974] and PREM [Dziewonski and

Anderson, 1981] the $n = -1$ mode seems to respond more to the difference in gross physical properties between the inner and outer cores. On the other hand PREM is almost neutral and the $n = -2$ modes and higher are at very long periods (as are these modes for a neutral core).

With the addition of rotation, and so far this has been carried out only to the inclusion of the next highest degree $l = 2$, the periods become severely constrained to lie below a projected limit of 13.3 hrs (approximately, for $m = 1$) shown as a dotted line in Figure 9. With further terms in the coupling chain included it is then probable that all such motions will have periods less than 12 hr. Hence it seems unlikely that a 13.9 hr. inner core period is possible unless the results for a truncated spherical harmonic expansion are invalid above 12 hrs (this is a distinct possibility for the inertia-gravity wave calculations).

A further possibility is that there is an interaction between the inner core motion and inertial waves. Certainly this would not be the case in a non-rotating Earth where the inner core translation is purely ${}_nS_1^m$. However in a rotating system, even for a neutrally-stratified fluid, toroidal displacements in the core (inertial waves) are coupled to spheroidal motions, excluding the elastic normal modes with positive radial number n, as discussed in Crossley and Rochester [1980]. Therefore the possibility coupling between the $n < 0$ motions described here and inertial waves must be allowed for in any complete dynamical description of core motions.

We conclude by recalling the comment by Smith [1976] that the $n = -1$ inner core motion should provide a useful constraint on the density jump at the inner core - outer core boundary. Model B1 has a density jump of 0.17 $gm\ cm^{-3}$ and PREM has 0.60.

Acknowledgments. We would like to thank Michael Rochester for discussions on the inertial wave identification. One of use (DJC) acknowledges support from NSERC (Canada) through Operating Grant # A4240.

References

Aldridge, K. D., Axisymmetric inertial oscillations of a fluid in a rotating spherical shell, *Mathematika*, *19*, 163-168, 1972.

Aldridge, K. D. and A. Toomre, Axisymmetric inertial oscillations of a fluid in a rotating spherical container, *J. Fluid Mech.*, *37(2)*, 306-323, 1969.

Aldridge, K. D. and L. I. Lumb, Inertial waves identified in the Earth's fluid outer core, *Nature*, *325*, 421-423, 1987.

Crossley, D. J., Core undertones with rotation, *Geophys. J. Roy. astr. Soc.*, *42*, 477-488, 1975.

Crossley, D. J., Oscillatory flow in the liquid core, *Phys. Earth Planet. Int.*, *36*, 1-16, 1984.

Crossley, D. J. and M. G.Rochester, Simple core undertones, *Geophys. J. Roy. astr. Soc.*, *60*, 129-161, 1980.

De Meyer, F., A multi-input-single output model for Earth tides data, *Bull. Inf. Marees Terr.*, *88*, 5628-5675, 1982.

Dziewonski, A. M. and D. L. Anderson, Preliminary reference Earth model, *Phys. Earth Planet. Int.*, *25*, 297-356, 1981.

Friedlander, S. and W. L. Siegmann, Internal waves in a rotating stratified fluid in an arbitrary gravitational field, *Geophys. Astr. Fluid Dyn.*, *19*, 267-291, 1982.

Greenspan, H.P., *The Theory of Rotating Fluids*, Cambridge University Press, 328pp., 1969.

Jeffreys, H. and R. O. Vicente, The theory of nutation and the variation of latitude, *Mon. Not. Roy. astr. Soc.*, *117*, 142-161, 1957.

Jordan, T. H. and D. L. Anderson, Earth structure from free oscillations and travel times, *Geophys. J. Roy. astr. Soc.*, *36*, 411-459, 1974.

Kudlick, M. D., On transient motions in a contained, rotating fluid, Ph. D. thesis, M.I.T., 1966.

Longuet-Higgins, M.S., The eigenfunctions of Laplace's equation over a sphere, *Phil. Trans. Roy. Soc. London, Ser. A*, *262*, 511-607, 1968.

Melchior, P. J., Diurnal tides and the Earth's liquid core, *Geophys. J. Roy. astr. Soc.*, *12*, 15-21, 1966.

Melchior, P. J. and B. Ducarme, Detection of inertial gravity oscillations in the Earth's core with a superconducting gravimeter at Brussels, *Phys. Earth Planet. Int.*, *42*, 129-134, 1986.

Molodensky, M. S., The theory of nutations and diurnal Earth tides, in *IVme Symp. Int. Sur Les Marees Terrestres*, Comm. Obs. Royal Belg., No. 188, Série Géophys. *58* , 25-56, 1961.

Mullan, D. J., Earthquake excitation and the geodynamo, *Science*, *181*, 553-554, 1973.

Neuberg, J., J. Hinderer and W. Zürn, Stacking gravity tide observations in Central Europe for the retrieval of the complex eigenfrequency of the nearly diurnal free wobble, , (in press).

Smith, M. L., Translational inner core oscillations of a rotating, slightly elliptical Earth, *J. Geophys. Res.*, *81*, 3055-3065, 1976.

Won, I.J. and J. T. Kuo, Oscillation of the Earth's inner core and its relation to the generation of the geomagnetic field, *J. Geophys. Res.*, *78*, 905-911, 1973.

Xi, Qinwen, The algebraic deduction of harmonic development for the tide-generating potential with the IBM-PC, in *Proceedings of the Tenth International Symposium on Earth Tides*, 481-489, Consejo Superior de Investigationes Cientificas, Madrid, 1986.

Zürn, W., B. Richter, P. A. Rydelek and J. Neuberg, Comment on : "Detection of inertial gravity oscillations in the Earth's core with a superconducting gravimeter at Brussels", *Phys. Earth Planet. Int.*, *49*, 176-178, 1987.

INERTIAL MODES IN THE EARTH'S FLUID OUTER CORE

Keith D. Aldridge, L. Ian Lumb and Gary A. Henderson

Department of Earth and Atmospheric Science, York University,
North York, Ontario, M3J 1P3, Canada

Abstract. Several inertial waves previously identified by Aldridge and Lumb [1987] in the long period gravimetric data of Melchior and Ducarme [1986], have been observed in a record from the same laboratory over 1440 days during the period 1982 to 1986. In the original data, collected following the Mindanao (20 November 1984) and Hindu Kush (30 December 1983) earthquakes, 8 inertial waves were tentatively identified by close proximity of their frequencies to those of a Poincaré Earth model for the fluid outer core. All of the waves identified in the earlier record persist in the longer record, thus verifying the significance of the earlier identification. All the waves decay more rapidly than would be expected from Ekman dissipation which suggests other mechanisms, possibly of hydromagnetic origin, for the damping of these waves. Alternatively the waves may be critically damped due to irregularity in the core-mantle boundary (CMB). Both of these models and others for the decay of the observed waves can be directly evaluated using the high-quality gravimetric data which is now available. The inertial waves identified here will serve as a precise tool for subsequent evaluation of models for core dynamics and the geodynamo.

Introduction

The gravimetric observations of Melchior and Ducarme [1986] (hereinafter MD) provide what appear to be the first data on the Earth's fluid core which will ultimately lead to an understanding of the core's structure and dynamics. This paper reports on the simplest possible Earth model which can be used to interpret their data: the fluid outer core is assumed to be a rigidly contained, incompressible, rotating fluid in adiabatic equilibrium. Although this model, named Poincaré after the governing equation describing it, does not provide any new information on core properties, it reveals the *primary* importance of rotation. Differences between observed frequencies and those interpreted as inertial waves from the Poincaré model will ultimately provide much needed information on the buoyancy, compressibility and departure from uniform rotation of the fluid outer core.

A review of the observations of MD is presented along with a discussion of the processing of their data. The Poincaré model is reviewed and its relationship to the subseismic wave equation of Smylie and Rochester [1981] (hereinafter SR) is described. Evidence for the existence of inertial waves in a rotating fluid contained in a spherical cavity comes from experimental observations and the properties of these waves are presented. Direct identification of some low-order modes in the data of MD is shown by comparison of their observed frequencies with those of the Poincaré model. Some discussion of possible excitation and damping mechanisms is given including the probable effect of irregular boundaries on the existence and damping of inertial waves.

Gravimetric Data

Spectra of gravimetric data have been presented by MD following the Hindu Kush (their Figure 1) and Mindanao (their Figure 2) earthquakes. In both of these figures two important properties are identifiable: a large amplitude spectral peak decreases with time since the earthquakes occurred and also *changes its frequency* over the same interval. The decay of this peak has an e-folding time of about 280 days; the same time scale of decay is seen in the change of the center frequency of this peak which appears to be asymptotic to about 13.9 hours. The similarity of the two time scales suggests that the decay of amplitude and change in frequency may be related through the same physical process and this is discussed below. The existence of a frequency which changes in time implies a non-stationary time series and accordingly great caution must be used in the application of significance tests to power spectra of this data. Recently, Zürn et al. [1987] reported that a record from their own instrument covering the same interval of time as part of the data presented by MD showed no significant power near the peak reported by MD. The apparent lack of stationarity of the data does not appear to have been considered by Zürn et al. [1987].

In addition to the major peak just described, there are a number of smaller ones, some of which clearly change their frequencies with time and remain visible in the longest records. These events will be discussed below.

Core Models

Currently the most comprehensive model of the Earth's fluid outer core is that described by the subseismic wave equation given by SR. There the core is modeled as a rotating, compressible, stratified, self-gravitating fluid contained by rigid boundaries. (Recently, Smylie [1987] has extended the subseismic description to allow for the elasticity of the mantle.) Their equation for the reduced pressure, χ, is given as:

$$L\chi = \sigma^2\nabla^2\chi - \frac{\partial^2\chi}{\partial z^2} - \frac{A}{B}\mathbf{C}\cdot\nabla\chi - \mathbf{C}^*\cdot\nabla\left(\frac{\mathbf{C}\cdot\nabla\chi}{B}\right) = 0 \quad (1)$$

where

$$A = \frac{\omega^2}{\beta}(\sigma^2 - 1) + \sigma^2(4\pi G\rho_0 - 2\Omega^2) + (\hat{\mathbf{k}}\cdot\nabla)\hat{\mathbf{k}}\cdot\mathbf{g}_0$$

$$B = \frac{\alpha^2\omega^2}{\beta}(\sigma^2 - 1) + \sigma^2 g_0^2 - (\hat{\mathbf{k}}\cdot\mathbf{g}_0)^2$$

$$\mathbf{C} = (\hat{\mathbf{k}}\cdot\mathbf{g}_0)\hat{\mathbf{k}} + i\sigma\hat{\mathbf{k}}\times\mathbf{g}_0 - \sigma^2\mathbf{g}_0$$

and

$$\sigma = \frac{\omega}{2\Omega}.$$

The coordinate z is in the direction $\hat{\mathbf{k}}$ which coincides with that of the rotation axis of Ω. The unperturbed density field is ρ_0 and g_0 is the magnitude of the unperturbed gravitational field, $\mathbf{g}_0$, while G is the gravitational constant. The adiabatic sound speed is α and ω is the wave frequency.

In the above description, the two scalar fields, χ and ψ, are defined as

$$\chi = \frac{p_1}{\rho_0} - V_1 \quad (2)$$

and

$$\psi = \beta\nabla\cdot\mathbf{u}, \quad (3)$$

where the subscript 1 refers to perturbation fields.

In the equation for ψ, β is the stratification parameter viz.,

$$\beta = 1 - \frac{\alpha^2}{\rho_0 g_0^2}\mathbf{g}_0\cdot\nabla\rho_0, \quad (4)$$

related to the Brunt-Väisälä frequency, N^2, via

$$N^2 = -\beta\left(\frac{g_0}{2\alpha\Omega}\right)^2, \quad (5)$$

which specifies the structure of the core.

The 'classical' Poincaré equation, which describes the contained motions of a homogeneous, incompressible contained rotating fluid may be obtained from 1. In the absence of stratification, ($\beta \longrightarrow 0$), the last term of equation 1 vanishes since $B \longrightarrow \infty$ and when $\alpha \longrightarrow \infty$, the second last term vanishes to give

$$\sigma^2\nabla^2\chi - \frac{\partial^2\chi}{\partial z^2} = 0. \quad (6)$$

The connection between these two fluid descriptions is now obvious and suggests the possibility of 'tracking' the laboratory inertial wave studies to the 'subseismically' defined Earth [Smylie and Rochester, 1986]. The effects of dissipation are neglected in the subseismic wave equation as well as the Poincaré equation although these are readily included in the latter case by the addition of a viscous boundary layer [e.g. Greenspan, 1968]. In fact consideration of dissipative effects is obviously essential to the interpretation of the experimental data on inertial waves. In what follows the existence of inertial waves in spherical shell geometry is reviewed and some properties of these waves are discussed.

Laboratory Observations of Inertial Waves

It has been established by several experimental studies that inertial waves exist in a rotating fluid contained by rigid spherical boundaries. The work described by Lumb and Aldridge (this issue, hereinafter LA) on the excitation of these waves with azimuthal wavenumber one in a spheroidal shell of rotating fluid is the most recent in a series of such studies which date back to the work of Aldridge and Toomre [1969] (hereinafter AT) who studied axially symmetric inertial waves of a fluid in a rotating spherical cavity. In the latter work the waves were excited by a reciprocating movement of the boundary and pressure measurements made while this occurred showed a portion of a spectrum (see Figure 3 of AT) of modes all with periods greater than half of the rotation period as predicted by linear theory from the solution of equation 6 with boundary condition that the normal component of the fluid velocity vanished. The spatial structure of some of the low-order axisymmetric inertial waves is shown in Figure 1 of AT.

The experimentally observed frequencies agree with those predicted from linear theory to within $\frac{1}{2}$ of 1% and the waves all decay with the spin-up time scale, $T = a/(\nu\Omega)^{1/2}$ where ν is the kinematic viscosity of the fluid, a is the radius of the sphere and Ω is rate of rotation (see AT). Some questions about the existence of inertial wave solutions have arisen for a fluid contained in a spheroidal shell of fluid because of the ill-posed aspect of the Poincaré problem and this has been discussed by Stewartson and Rickard [1970], Stewartson and Walton [1976] and Stewartson [1978]. Existence of these waves in a real (viscous) fluid has been verified by Aldridge [1972, 1975] (hereinafter A1, A2) for the axisymmetric case and more recently by Stergiopoulos and Aldridge [1984] (hereinafter SA) and LA for the case of azimuthal wavenumber one. The absence of closed form solutions in the shell geometry has limited the interpretation of these experiments to approximate solutions of variational type and even these solutions exhibit what appears to be classic asymptotic divergence [A1, A2], presumably related to the ill-posed aspect of this problem.

The most recent experimental work reported in this issue by LA verifies the existence of several spatially simple inertial waves with azimuthal wavenumber one in a spheroidal shell of rotating fluid . The waves can travel in either the same sense as the rotation (prograde) or opposite to the rotation (retrograde) and they exhibit interaction during their free decay which occurs on the spin-up time scale T referred to above. The measured eigenfrequencies and decay rates for several of these modes for various excitation amplitudes are given in LA as well as a tentative identification through comparison with predicted eigenfrequencies for a fluid sphere. Previous experimental work reported in A2 showed that measured eigenfrequencies of similar spatial complexity differed by at most a few percent from those for a sphere due to the presence of a similarly sized inner body. Schematics for instantaneous velocity fields for a few of these are shown in Figure 1.

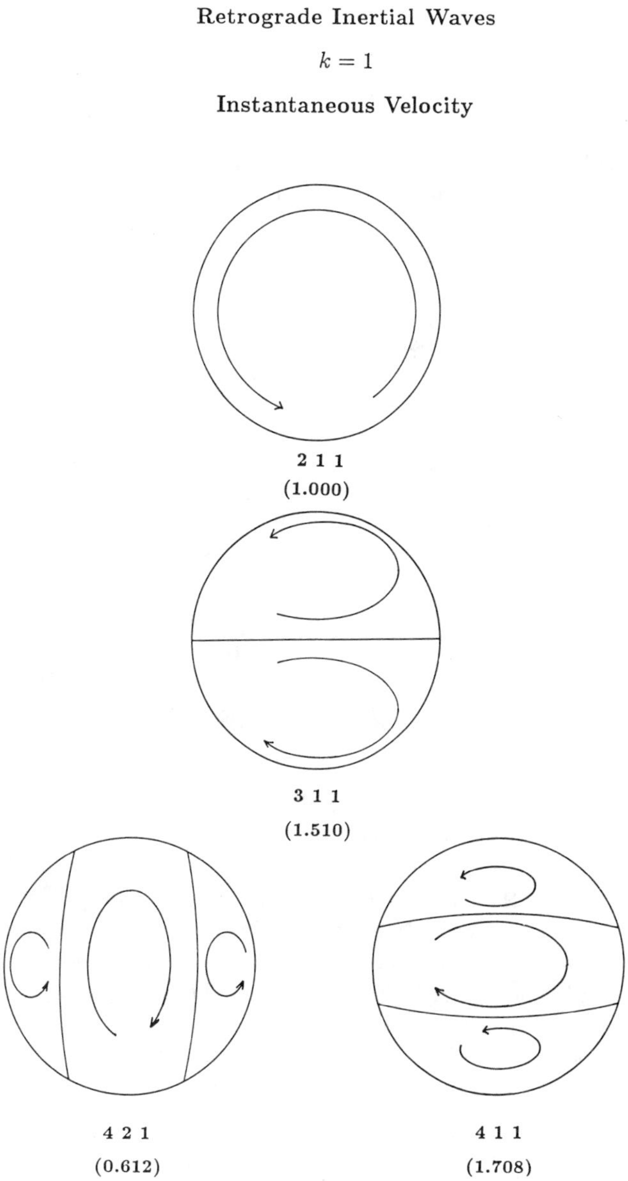

Fig. 1. Schematics of instantaneous velocity fields for some low-order, azimuthal number unity, retrograde inertial modes where (n, m, k) is the modal numbering scheme (described in text) and the number in parentheses is the associated eigenvalue of the mode.

Identification of Inertial Waves in the Gravimetric Data

An identification of several inertial waves in the gravimetric data of MD has been carried out by Aldridge and Lumb [1987], (hereinafter AL). A comparison of observed periods measured by MD was made with those predicted from a period equation closely related to that derived from the Poincaré equation 6 but in addition accounting for the slight ellipticity of the CMB. Solution of the the Poincaré equation in a spheroidal cavity [Kudlick, 1966] gives the inertial wave periods $T_{nmk} = 2\,\pi\;/\;\omega_{nmk}$ from the m roots

$$x_{nmk} = \frac{\omega_{nmk}/2\Omega}{[1+\epsilon(1-(\omega_{nmk}/2\Omega)^2)]^{1/2}}$$

of

$$(1-x^2)\,\frac{dP_n^k(x)}{dx} = k\left(\frac{1+\epsilon x^2}{1+\epsilon}\right)^{1/2} P_n^k(x), \qquad (7)$$

where Ω is the Earth's rotation rate and $\epsilon = 1/(1-f)^2-1$ and $f = 1/392$. For each n of the Legendre Polynomial P_n^k, the smallest value of m corresponds to the simplest radial structure and k is the azimuthal wavenumber. The calculated periods and modal numbers are given in columns three and four respectively of Table 1.

TABLE 1. Comparison of Periods Observed by Melchior and Ducarme with Calculated Inertial Wave Periods (after AL).

Observed period (hrs)		Calculated period (hrs)	mode (n,m,k)
Mindanao	Hindu Kush	T_{nmk}	
15.3	15.2	15.832	(3,1,1)
13.9	14.0	14.003	(4,1,1)
13.2	13.2	13.247	(5,1,1)
—	15.6	15.626	(5,1,0)
14.5	14.3	14.403	(6,1,0)
13.6	13.6	13.720	(7,1,0)
13.4	13.4	13.290	(8,1,0)

Agreement between the observed and calculated periods for inertial waves with $k = 1$, shown in the upper part of the table, is within about 3%. It is important to point out that the calculations given here do not include effects of the inner core. This apparently minor correction has proven to be difficult to obtain due to the ill-posedness of the boundary value problem as pointed out above. Experimental measurements of the eigenfrequencies of inertial waves in spherical and spheroidal shells have been made by SA and LA, however, and they differ from those of the sphere by only a few percent for the spatially simple modes shown in Table 1. It appears, therefore, that the data of MD can be interpreted as the discovery of predominantly inertial waves in the Earth's fluid outer core.

It is important to note that the observed wave periods in Table 1 are estimates of what appear to be limiting values in the data of MD. As noted above these periods change over time during the decay of amplitude and accordingly it was decided that best estimates of the ultimate values would be the most suitable for comparison with those from the Poincaré model. More recent results presented at the meeting by Melchior et al. [1987] showed the same type of time dependence during the time that the amplitude decayed.

The reasons for expecting inertial waves of azimuthal wavenumbers 0, 1 and 2 have been given earlier by AL. In summary, it would be expected to find k=1,2 if the origin of the excitation is an earthquake since this is the expected form of the displacement field [e.g. Mansinha et al., 1979], while k=0 type waves would be excited by a change in rotation speed which has been inferred from the change in observed frequency with time for the larger amplitude waves. The latter point follows directly from the fact that inertial wave periods depend only on rotation speed and the container's geometry which is expected to be essentially unchanged during the free decay of the waves.

It is important to note that the observations of MD cover only a small portion of the inertial wave spectrum. Even though there is a dense spectrum of inertial waves, the ones identified in the upper portion of Table 1 correspond to the only modes for k=1 in this range of period. All other modes of the same spatial complexity (n=3,4,5) lie outside this range.

Graphical illustration is a convenient way to make this point. Plotted in Figure 2 are the data of MD in dimensionless form. Their observed frequencies, from a record including the Hindu Kush and Mindanao events and extending over 1440 days, have been divided by the rotation rate of the Earth and are shown as solid bars whose amplitudes are proportional to spectral power as supplied by Dehant [personal communication, 1987]. Plotted on the same axis in dashed lines are the eigenfrequencies for inertial waves with $n = 3, 4, \ldots, 7$. It would be surprising to excite those waves of much higher spatial complexity by a simple spatial perturbation as would be expected from an earthquake; it would be even more surprising to detect such a wave because of the high degree of self-cancellation expected in the velocity field for such modes. These remarks follow directly from the experience of excitation and detection of inertial waves in the laboratory by AT, A1, A2, SA and LA. The theoretical frequencies all correspond to inertial waves propagating in the retrograde sense. The close proximity of the (311), (411) and (511) modes already seen in Table 1 is evident in this figure. The only mode with k=1 and simpler spatial structure is the (211) mode and it is clearly outside the frequency range of the observed modes. The same is true for other members of the n= 3, 4, 5 series shown in this figure.

The data plotted in Figure 2 has been replotted (solid lines) in Figure 3 where λ is less than zero so that it can be compared to the frequencies which correspond to inertial waves (dashed lines) traveling in the prograde direction. Here we note that there is of course no way to tell which way the observed waves are propagating from observations at one location, other than by comparison with a predicted frequency that signifies a unique direc-

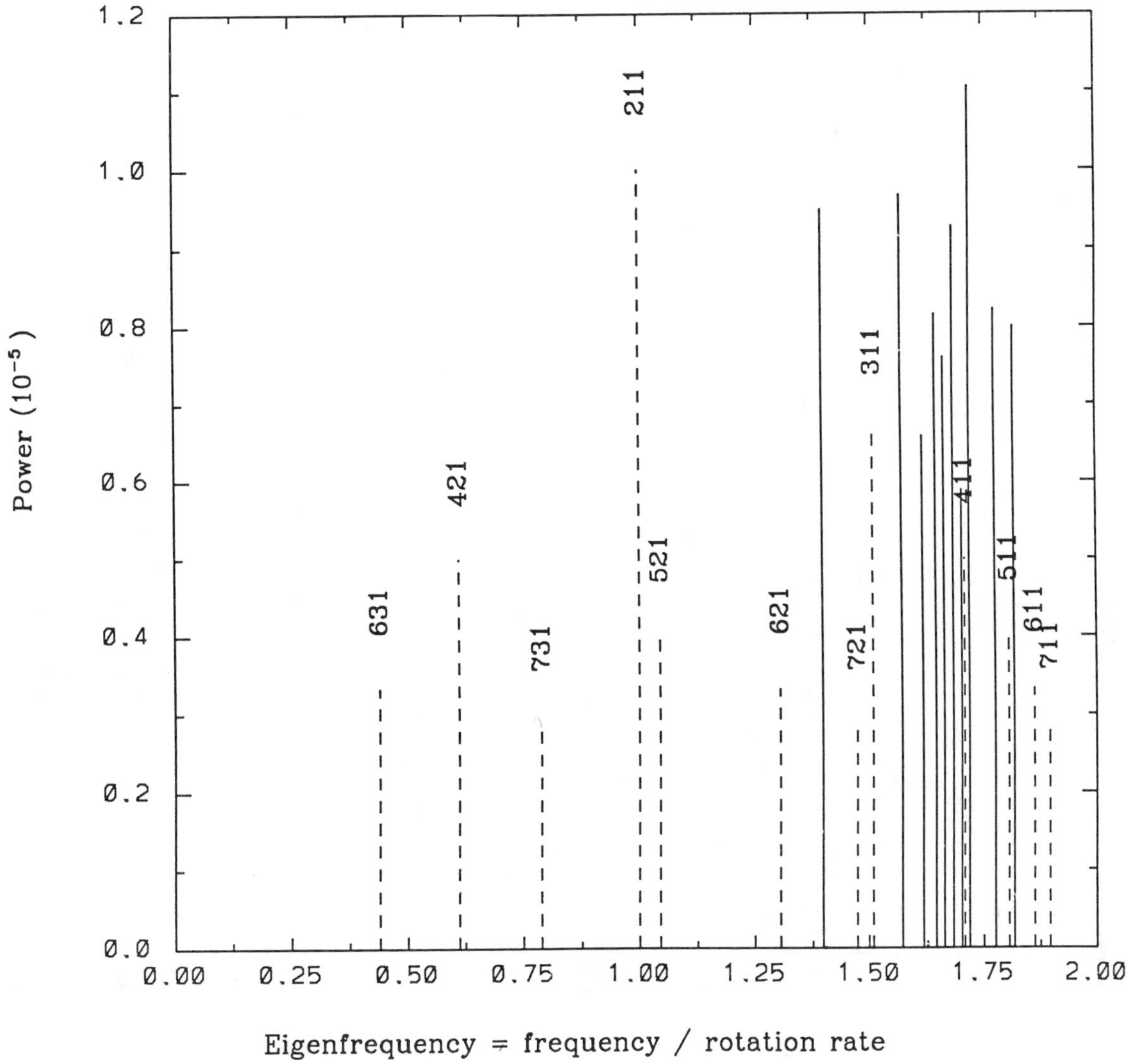

Fig. 2. Power spectrum of superconducting gravimetric data (solid lines) over a 1440 day period [*Dehant,* personal communication, 1987] superimposed with low-order, azimuthal number unity, retrograde inertial waves (dashed lines).

tion. The predicted frequencies for n = 3, 4, 5 are all outside the range of observed frequencies thus confirming the identification in Table 1: of the ten waves with k =1, for the cases n =2 (1), 3 (2), 4 (3), 5 (4) which are those with the simplest possible spatial structure, the three which fall within the range of observed frequencies coincide with those predicted from the Poincaré model. The discrepancy between the largest observed mode and the (411) is only a fraction of a percent, that of the (511) is slightly greater while for the (311) the difference is about 3%.

Some of the low spatial order axisymmetric modes are plotted (dashed lines) in Figure 4 along with the dimensionless eigenfrequencies from Dehant [personal communication, 1987]. The (510), (610), (710) and (810) all lie within the range of the observed frequencies although identification of these with measured counterparts is difficult to make in this figure which only shows a qualitative comparison between model and observation.

Figures 2, 3 and 4 are presented to show how the identification process was carried out in Table 1 since there are indeed an infinite number of eigenmodes in the iner-

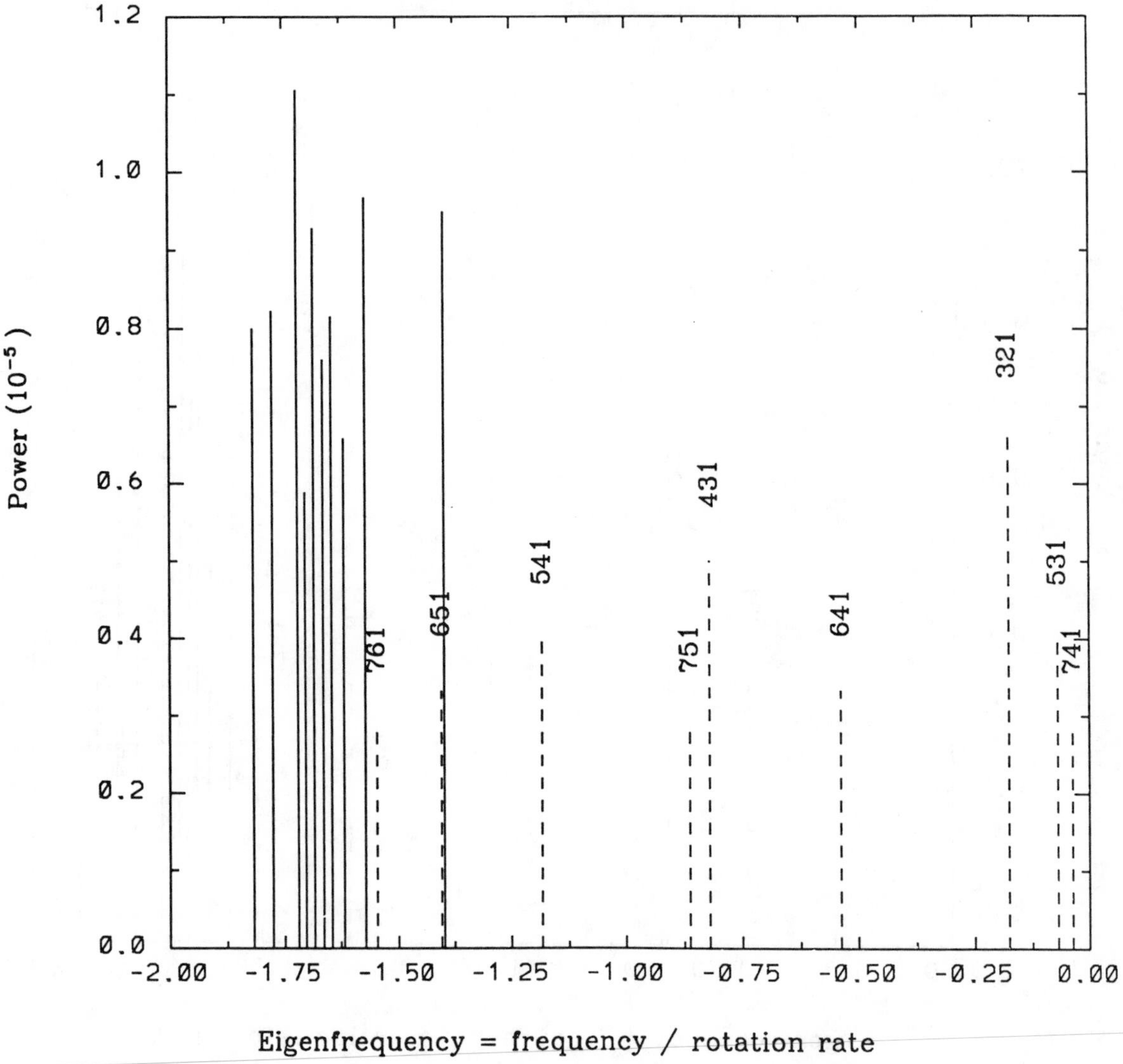

Fig. 3. As Figure 2 except that the dashed lines represent prograde inertial waves.

tial wave spectrum. There seems little reason to add the higher order modes in these figures because physical reasons given above argue against the existence and detection of such modes.

Irregular Boundaries

The geophysical motivation for considering irregular boundaries dates back at least to Hide's [1967] suggestion of 'bumps' on the CMB. Recent results from a variety of different sources has once again focussed attention in this area. The seismic tomographic inversion of body wave data is suggestive of 'mountains' of ± 6 km [Morelli and Dziewonski, 1987] or possibly 'continents' of comparable relief [Creager and Jordan, 1986]. Hager et. al. [1985] used density contrasts in the lower mantle, which inferred the long-wavelength features of the geoid from seismic tomography, to arrive at ± 3 km relief of dynamically maintained topography at the CMB. (According to Gubbins and Richards [1986], there are two components of CMB topography: dynamic, resulting from buoyancy-driven flow in the mantle; and isostatic, resulting from chemically stable heterogeneities.) Similar results to Hager et. al.'s work is, in principle, available from the more recent work of Forte and Peltier [1987]. Malin and Hide [1982] revisited their earlier work on correlations between the Earth's gravitational and geomagnetic fields while other workers have studied the effects CMB topog-

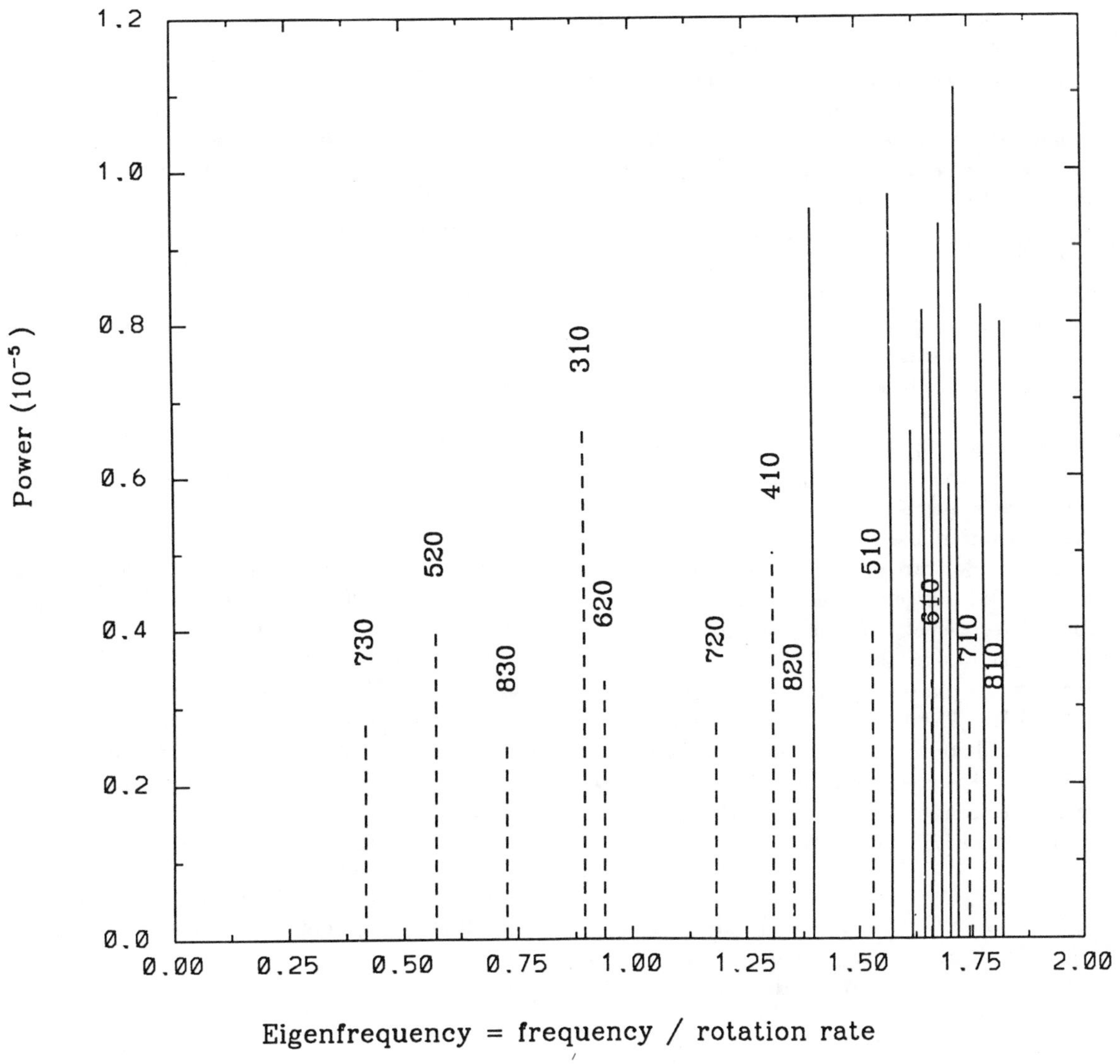

Fig. 4. As Figure 2 except that the dashed lines represent standing inertial waves.

raphy on the magnetic field [Gubbins and Richards, 1986; Hide, 1987]. Lately, CMB irregularities have captured the interests of mantle dynamicists [Davies and Gurnis, 1986; Zhang and Yuen, 1986] who are involved in modeling the lateral heterogeneity of the D'' region. Further observational results from very-long-baseline interferometry (VLBI) [e.g. Gwinn et. al., 1986] hold the promise that the effects of CMB relief on the rotational dynamics may be measured. Although it is obvious that this problem has been approached from several different 'angles', as Aldridge and Lumb [1987] indicate, it is important to investigate the effect of CMB topography on inertial waves in the fluid outer core.

Reflection of inertial waves from an irregular surface can result in critical damping of the waves as shown by Phillips [1963]. Since the CMB is likely to appear *rough* on a scale of wavelengths comparable to the core's radius, some concern for this phenomenon is appropriate. Here we consider the problem of existence of inertial waves in a geometry where trapping is possible.

The simplest possible model for the study of such damping is that of a conical boundary as studied experimentally by Beardsley [1970]. There it was found that no inertial wave spectrum of the type studied by AT, A1, A2, SA and LA could be obtained by trying to excite axisymmetric inertial waves in a cone of fluid rotating about its axis of symmetry. When a small insert was included in the end of the cone to produce a frustum, a discrete spectrum was observed. It is the purpose here to show how this result can be interpreted.

It was understood by Beardsley that critical damping took place in a cone because wave energy was repeatedly reflected into the tip of the cone so that no reinforcement of wave energy was possible. The situation was seen to be analogous to a string stretched between supports but with one of the supports removed to infinity. Thus wave energy never returns once it is sent down the string. When the cone is converted to a frustum, reflection at the base can occur and a spectrum is observed.

No quantitative interpretation of the measured eigenfrequencies was made by Beardsley except to note that by considering the frustum to be a perturbed cone, no estimates of eigenfrequencies were possible.

At present a variational principle has been developed and applied to the Poincaré problem for a conical frustum by Henderson [1988]. Eigenfrequencies have been calculated from a finite element implementation and some of the results are shown in Table 2. The bottom line of the table shows the dimensionless eigenfrequencies measured by Beardsley. The first column gives the number of mesh elements while the remaining columns show the calculated dimensionless eigenfrequencies for each of three resonances. Agreement between the finite element results and observations is within about 3%.

TABLE 2. Comparison of Measured Axisymmetric Eigenfrequencies from Beardsley [1970] with those from the Finite Element Calculation by Henderson [1988].

Number of elements	Principal resonances[1] (1,0,2)	(1,0,3)	(1,0,4)
5	1.1723	1.3821	1.5597
33	1.1219	1.3310	1.5141
49	1.1165	1.3252	1.5098
72	1.1190	1.3274	1.5160
Measured	1.12	1.30	1.48

[1]The modal numbering scheme used here, (n, k, m), is that of Henderson's, [1988], where n is the degree of the Bessel function in the associated period equation for the cylinder [Greenspan, 1968], k is the azimuthal wavenumber, and m is the mth root of this period equation.

The above method will be applied to other irregular boundaries. The possibility of inertial wave trapping can then be investigated for the more general case in the fluid outer core.

Conclusions

Properties of the Earth's fluid outer core appear to be accessible through gravimetric observations at long periods. In particular, low-order inertial waves which depend entirely on rotation can account for most of the observed spectrum presented by MD. Small discrepancies between observed frequencies and those of the Poincaré model are presumably due to buoyancy, compressibility and nonuniform rotation of the fluid core which are neglected in this model. Dissipation is not included in this discussion. Decay rates calculated from the data of MD seem too large to be explained by any model based on Ekman dissipation even with an enhanced eddy viscosity. It would appear more likely that electromagnetic damping is important in a CMB layer and that the small changes in eigenfrequency observed during the decay of the waves are closely related to this effect.

More gravimetric data, from a global network of instruments, is needed to extend the above interpretations. Further evidence for core oscillations at inertial wave periods will come from the residuals in VLBI data [e.g. Herring et al., 1986] at these periods. Independent confirmation of the existence of core modes will provide the needed observational evidence for the full development of core models and the ultimate measurement of core properties.

Acknowledgements. We thank Doug Smylie and Tony Szeto for several helpful discussions. This research was supported by a generous operating grant from NSERC.

References

Aldridge, K. D., Axisymmetric inertial oscillations of a fluid in a rotating spherical shell, *Mathematika, 19,* 163-168, 1972.

Aldridge, K. D., Inertial waves and the Earth's outer core, *Geophys. J. R. astr. Soc., 42,* 337-345, 1975.

Aldridge, K. D. and L. I. Lumb, Inertial waves identified in the Earth's fluid outer core, *Nature, 325,* 421-423, 1987.

Aldridge, K. D. and A. Toomre, Axisymmetric inertial oscillations of a fluid in a rotating spherical container, *J. Fluid Mech., 37,* 307-323, 1969.

Beardsley, R. C., An experimental study of inertial waves in a closed cone, *Stud. Appl. Math, 49,* 187-196, 1970.

Creager, K. C. and T. H. Jordan, Aspherical structure of the core-mantle boundary from *PKP* travel times, *Geophys. Res. Lett., 13,* 1497-1500, 1986.

Davies, G. F. and M. Gurnis, Interaction of mantle dregs with convection: lateral heterogeneity at the core-mantle boundary, *Geophys. Res. Lett., 13,* 1517-1520, 1986.

Forte, A. M. and W. R. Peltier, Plate tectonics and aspherical Earth structure: The importance of poloidal-toroidal coupling, *J. geophys. Res., 92,* 3645-3679, 1987.

Greenspan, H. P., *The Theory of Rotating Fluids,* Cambridge University Press, 328 pp., 1969.

Gubbins, D. and M. Richards, Coupling of the core dynamo and mantle: thermal or topographic?, *Geophys. Res. Lett., 13,* 1521-1524, 1986.

Gunn, J. S. and K. D. Aldridge, Inertial waves in a non-uniformly rotating fluid, (this issue).

Gwinn, C. R., T. A. Herring and I. I. Shapiro, Geodesy by radio interferometry: Studies of the forced nutations of the Earth 2. Interpretation, *J. geophys. Res., 91,* 4755-4765, 1986.

Hager, B. H., R. W. Clayton, M. A. Richards, R. P. Comer and A. M. Dziewonski, Lower mantle heterogeneity, dynamic topography and the geoid, *Nature, 313,* 541-545, 1985.

Henderson, G. A., *M. Sc. thesis,* York University, (in preparation).

Herring, T. A., C. R. Gwinn and I. I. Shapiro, Geodesy by radio interferometry: Studies of the forced nutations of the Earth 1. Data analysis, *J. geophys. Res., 91,* 4745-4754, 1986.

Hide, R. Motions of the Earth's core and mantle and variations of the main geomagnetic field, *Science, 157,* 55-56, 1967.

Hide, R., Motions in the Earth's core and mantle and variations of the main geomagnetic field revisited, (this issue).

Kudlick, M. D., On transient motions in a contained, rotating fluid, *Ph.D. thesis,* M.I.T., 1966.

Lumb, L. I. and K. D. Aldridge, An experimental study of inertial waves in a spheroidal shell of rotating fluid, (this issue).

Malin, S. R. C. and R. Hide, Bumps on the core-mantle boundary: geomagnetic and gravitational evidence revisted, *Phil. Trans. R. Soc. Lond. A306,* 281-289, 1982.

Mansinha, L., D. E. Smylie and C. H. Chapman, Seismic excitation of the Chandler wobble revisited, *Geophys. J. R. astr. Soc., 59,* 1-17, 1979.

Melchior, P. W., D. J. Crossley, V. P. Dehant and B. Ducarme, Have inertial waves been identified from the Earth's core?, (this issue).

Melchior, P. and B. Ducarme, Detection of inertial gravity oscillations in the Earth's core with a superconducting gravimeter at Brussels, *Phys. Earth planet. Int., 42,* 129-134, 1986.

Morelli, A. and A. M. Dziewonski, Topography of the core-mantle boundary and lateral homogeneity of the liquid core, *Nature, 325,* 678-683, 1987.

Phillips, O. M., Energy transfer in rotating fluids by reflection of inertial waves, *Phys. Fluids, 6,* 513-520, 1963.

Smylie, D. E., Variational calculation of wobble modes of the Earth (this issue).

Smylie, D. E. and M. G. Rochester, Compressibility, core dynamics and the subseismic wave equation, *Phys. Earth planet. Int., 24,* 308-318, 1981.

Smylie, D. E. and M. G. Rochester, A variational principle for the subseismic wave equation, *Geophys. J. R. astr. Soc., 86,* 553-561, 1986.

Stergiopoulos, S. and K. D. Aldridge, Ringdown of inertial waves in a spheroidal shell of rotating fluid, *Phys. Earth planet. Int., 36,* 17-26, 1984.

Stewartson, K., Homogeneous fluids in rotation. Section B: Waves in *Rotating Fluids in Geophysics,* (eds.) P. H. Roberts and A. M. Soward, Academic Press, 67-103, 1978.

Stewartson, K. and J. A. Rickard, Pathological oscillations of a rotating fluid, *J. Fluid Mech., 35,* 759-773, 1969.

Stewartson, K. and I. C. Walton, On waves in a thin shell of stratified rotating fluid, *Proc. R. Soc. Lond.* A*349,* 141-146, 1976.

Zhang, S. and D. A. Yuen, Deformation of the core-mantle boundary induced by spherical-shell, compressible convection, *Geophys. Res. Lett., 13,* 899-902, 1986.

Zürn, W., Richter, B., Rydelek, P.A. & Neuberg, J. (1987) Comment on 'Detection of inertial gravity oscillations in the Earth's core with a superconducting gravimeter at Brussels' *Phys. Earth planet. Int., 49,* 176-178.

VARIATIONAL CALCULATION OF CORE MODES IN REALISTIC EARTH MODELS

D. E. Smylie

Department of Earth and Atmospheric Science, York University,
North York, Ontario M3J 1P3 Canada

Abstract. A variational procedure for calculating the eigenfrequencies and eigenfunctions of long period core modes ($< 300\mu$Hz) is constructed and implemented. For the first time, realistic Earth properties are fully taken into account, including compressibility and stratification of the fluid core, self-gravitation, and elasticity of the shell. Combined with the possible recent gravimetric observation of such modes, a new field of long period core spectroscopy may be emerging.

Introduction

Study of the long period oscillations of rotating, self-gravitating fluid masses has its roots in the classical literature. Work done in the classical period culminated in the development of the governing equation (Poincaré, 1885) and its solution for spheroids of finite ellipticity by Bryan (1889).

The subject of rotating fluids experienced a revival in the 1960's with the beautiful laboratory experiments of Aldridge and Toomre (1969) and the publication of a monograph on the subject by Greenspan (1969). Although the laboratory experiments confirmed the classical theory, certain theoretical difficulties emerge when, as in the case of the Earth's core, a central body is present (Stewartson and Rickard, 1969). Curiously, these difficulties are not apparent in spherical shell experiments (Aldridge, 1972).

In addition to the presence of the solid inner core, the real Earth differs in several essential ways from the assumptions of classical theory and the conditions of laboratory experiments, where the fluid is considered incompressible and of uniform density, and to be contained in a rigid envelope. A realistic Earth model allows for the elasticity of the shell, stratification and compressibility of the outer fluid core, and self-gravitation.

The possible recent gravimetric observation of oscillations of the fluid core (Melchior and Ducarme, 1985), and their identification with some fundamental inertial modes (Aldridge and Lumb, 1987), raises the hope that with careful account of the realistic properties of the Earth, a new spectroscopic tool may be developed for studies of the core.

Core modes are also of importance in understanding the rotational dynamics of the Earth. Modes of azimuthal number unity exchange angular momentum with the shell and are therefore reflected in the wobble, length of day and nutation spectra. With the vastly increased precision of new rotation observational techniques such as VLBI, it may be possible, with a good theoretical model for the rotational dynamics of the Earth, to infer structural and thermal properties of the fluid core.

The purpose of the present paper is to set the foundation for such studies by constructing and implementing a variational procedure for the calculation of the eigenfrequencies and eigenfunctions of core modes in realistic Earth models. In a companion paper in this volume (Smylie, 1988), the variational method is adapted to the computation of wobble-nutation modes.

Variational Principles for the Earth

Small amplitude oscillations of the Earth are governed by the equation of motion

$$\rho_0\omega^2\mathbf{u} - 2i\rho_0\omega\mathbf{\Omega}\times\mathbf{u} = \mathbf{L}(\mathbf{u}, V_1) \tag{1}$$

and the Poisson gravitational equation

$$\nabla^2 V_1 = -4\pi G\rho_1, \tag{2}$$

where ρ_0 is the equilibrium mass density and ρ_1 its perturbation, ω is the angular frequency of the oscillation,

$\mathbf{u}$ is the vector displacement field, $\mathbf{\Omega}$ represents the uniform rotation of the reference frame about a fixed spatial direction at the mean rate of Earth's rotation, V_1 is the negative of the perturbation in gravitational potential and $\mathbf{L}$ is a real, linear vector operator giving the elasto-gravitational resistance to deformation.

In the fluid outer core

$$\mathbf{L}(\mathbf{u}, V_1) = \nabla p_1 - \rho_1 \mathbf{g}_0 - \rho_0 \nabla V_1 = \rho_0 \nabla \chi + (1-\eta)\rho_0 \mathbf{g}_0 \nabla \cdot \mathbf{u}. \quad (3)$$

Here η is the stratification parameter, expressing the ratio of the actual density lapse rate to the adiabatic lapse rate, and

$$\chi = \frac{p_1}{\rho_0} - V_1, \quad (4)$$

with p_1 the perturbation in pressure. $\mathbf{g}_0$ is the equilibrium gravity vector. Both p_1 and ρ_1 are measured at the undeformed coordinate points and are therefore calculated to include transport terms as

$$p_1 = -\rho_0 \alpha^2 \nabla \cdot \mathbf{u} - \rho_0 \mathbf{u} \cdot \mathbf{g}_0, \quad (5)$$

$$\rho_1 = -\rho_0 \nabla \cdot \mathbf{u} - \mathbf{u} \cdot \nabla \rho_0 = -\nabla \cdot (\rho_0 \mathbf{u}), \quad (6)$$

where α^2 is the square of the P-wave velocity.

Both the acceleration terms and the linear vector operator $\mathbf{L}$ in equation (1) are Hermitean. Hermiticity of the acceleration terms is easily established from the properties of the triple scalar product, while for $\mathbf{L}$, as given by (3) for the fluid outer core, application of the divergence theorem and some rearrangement of terms yields

$$\int \tilde{\mathbf{u}}^* \cdot \mathbf{L}(\mathbf{u}, V_1)\, d\mathcal{V} = \int p_1 \tilde{\mathbf{u}}^* \cdot \hat{\mathbf{n}}\, d\mathcal{S} - \int \rho_0 \tilde{\mathbf{u}}^* \cdot \nabla V_1 d\mathcal{V} + \int \rho_0 \alpha^2 (\nabla \cdot \tilde{\mathbf{u}}^*)(\nabla \cdot \mathbf{u})\, d\mathcal{V} + \int (\rho_0 \mathbf{u} \cdot \mathbf{g}_0 \nabla \cdot \tilde{\mathbf{u}}^* + \rho_0 \tilde{\mathbf{u}}^* \cdot \mathbf{g}_0 \nabla \cdot \mathbf{u})\, d\mathcal{V} + \int \eta \frac{\rho_0}{\alpha^2} (\tilde{\mathbf{u}}^* \cdot \mathbf{g}_0)(\mathbf{u} \cdot \mathbf{g}_0)\, d\mathcal{V}, \quad (7)$$

where $\tilde{\mathbf{u}}^*$, $\tilde{V}_1^*$ describe a deformation field that is independent of that described by $\mathbf{u}$, V_1. Using (2) and (6), the identity

$$-\int \rho_0 \tilde{\mathbf{u}}^* \cdot \nabla V_1 d\mathcal{V} = \frac{1}{4\pi G} \int \left[\nabla \cdot \left\{ V_1 \left(\nabla \tilde{V}_1^* - 4\pi G \rho_0 \tilde{\mathbf{u}}^* \right) \right\} - \nabla \tilde{V}_1^* \cdot \nabla V_1 \right] d\mathcal{V} \quad (8)$$

is easily established, which on substitution in (7), produces

$$\int \tilde{\mathbf{u}}^* \cdot \mathbf{L}(\mathbf{u}, V_1)\, d\mathcal{V} = \int p_1 \tilde{\mathbf{u}}^* \cdot \hat{\mathbf{n}}\, d\mathcal{S} + \frac{1}{4\pi G} \int V_1 \left(\nabla \tilde{V}_1^* - 4\pi G \rho_0 \tilde{\mathbf{u}}^* \right) \cdot \hat{\mathbf{n}}\, d\mathcal{S} - \frac{1}{4\pi G} \int \nabla \tilde{V}_1^* \cdot \nabla V_1 d\mathcal{V} + \int \rho_0 \alpha^2 (\nabla \cdot \tilde{\mathbf{u}}^*)(\nabla \cdot \mathbf{u})\, d\mathcal{V} + \int [\rho_0 \mathbf{u} \cdot \mathbf{g}_0 \nabla \cdot \tilde{\mathbf{u}}^* + \rho_0 \tilde{\mathbf{u}}^* \cdot \mathbf{g}_0 \nabla \cdot \mathbf{u}]\, d\mathcal{V} + \int \eta \frac{\rho_0}{\alpha^2} (\tilde{\mathbf{u}}^* \cdot \mathbf{g}_0)(\mathbf{u} \cdot \mathbf{g}_0)\, d\mathcal{V}. \quad (9)$$

For reasons of symmetry we then have

$$\int \tilde{\mathbf{u}}^* \cdot \mathbf{L}(\mathbf{u}, V_1)\, d\mathcal{V} - \int \mathbf{u} \cdot \mathbf{L}\left(\tilde{\mathbf{u}}^*, \tilde{V}_1^*\right) d\mathcal{V} = \int (p_1 \tilde{\mathbf{u}}^* - \tilde{p}_1^* \mathbf{u}) \cdot \hat{\mathbf{n}}\, d\mathcal{S} + \frac{1}{4\pi G} \int \left[V_1 \left(\nabla \tilde{V}_1^* - 4\pi G \rho_0 \tilde{\mathbf{u}}^* \right) - \tilde{V}_1^* (\nabla V_1 - 4\pi G \rho_0 \mathbf{u}) \right] \cdot \hat{\mathbf{n}}\, d\mathcal{S}. \quad (10)$$

Relation (10) also holds for the solid parts of the Earth (shell and inner core), being in essence the generalization of Betti's reciprocal theorem of elasticity to realistic Earth models (Smylie and Mansinha, 1971). At the free surface of the Earth, the integrals on the right of (10) vanish, while continuity of radial displacement, stress, gravitational potential and potential gradient ensure that those arising from the fluid core are cancelled by the surface integrals over its boundaries contributed by the solid inner core and the shell. We are thus provided with a Lagrangian functional for the whole Earth (Johnson and Smylie, 1977), which is stationary at eigenvalue-eigenfunction pairs, with the form

$$\mathcal{L}(\mathbf{u}, V_1) = \omega^2 \int \rho_0 \mathbf{u}^* \cdot \mathbf{u} d\mathcal{V} - 2\omega \int \rho_0 \mathbf{u}^* \cdot (i\mathbf{\Omega} \times \mathbf{u})\, d\mathcal{V} - \int \mathbf{u}^* \cdot \mathbf{L}(\mathbf{u}, V_1)\, d\mathcal{V}. \quad (11)$$

Owing to the strong elastic restoring forces in the solid parts of the Earth, the dynamical terms in the shell and inner core are small by comparison, and it is more convenient to solve for deformations there directly. Only the part of the functional (11) arising from the outer fluid core need therefore be made stationary, provided the trial functions are properly matched to the solutions in the solid parts of the Earth through the boundary conditions. Using the form on the right side of expression (3) for the operator $\mathbf{L}$ appropriate for the fluid core, we require stationarity only in the reduced functional

$$\mathcal{L}_{OC} = \omega^2 \int \rho_0 \mathbf{u}^* \cdot \mathbf{u} d\mathcal{V} - 2\omega \int \rho_0 \mathbf{u}^* \cdot (i\mathbf{\Omega} \times \mathbf{u})\, d\mathcal{V} - \int \rho_0 \mathbf{u}^* \cdot \nabla \chi d\mathcal{V} - \int (1-\eta) \rho_0 \mathbf{u}^* \cdot \mathbf{g}_0 \nabla \cdot \mathbf{u} d\mathcal{V}. \quad (12)$$

Specialization to the Fluid Core and the Subseismic Equation

Even at the relatively long periods contemplated here, for core scale wavelengths, the motion is highly isentropic. Significant departure from adiabaticity arises only when periods approaching convective time scales are reached. Scaling of a barotropic (adiabatic) equation of state, combined with the mass conservation equation, shows that to less than parts in 10^3 (Smylie et al, 1984), the flow, and for small harmonic motions, the displacement field, obey

$$\nabla \cdot \mathbf{u} + \frac{1}{\alpha^2}\mathbf{u} \cdot \mathbf{g}_0 = 0. \tag{13}$$

In physical terms, the approximation is to take into account, in calculating compression-dilatation, only the contribution of transport through the background hydrostatic pressure field. Consistent with the assumptions that the periods involved are long compared to those of acoustic modes but short compared to the time scales of convection, the effects on the density of perturbations in flow pressure and temperature are ignored. Since the frequencies for which the approximation is valid are below traditional seismic frequencies, we refer to it as the subseismic approximation.

Under this approximation, the acceleration terms on the left side of (1) may be replaced by

$$\rho_0 \left[\nabla \chi - \frac{1-\eta}{\alpha^2}(\mathbf{u} \cdot \mathbf{g}_0)\mathbf{g}_0\right], \tag{14}$$

leading to the functional

$$\mathcal{L}_{OC} = -\int \rho_0 (1-\eta)(\mathbf{u}^* \cdot \mathbf{g}_0)\left[\nabla \cdot \mathbf{u} + \frac{1}{\alpha^2}\mathbf{u} \cdot \mathbf{g}_0\right] d\mathcal{V}. \tag{15}$$

This functional is formed by multiplying through the left side of (13) by $-\rho_0(1-\eta)(\mathbf{u}^* \cdot \mathbf{g}_0)$ and integrating over the outer core. A functional which is more convenient to implement numerically can be obtained by multiplying through the left side of (13) with a different scalar.

As has been shown by Friedlander (1985), equation (13) may be reduced to the solenoidal form

$$\nabla \cdot (f\mathbf{u}) = 0, \tag{16}$$

provided the scalar f obeys

$$\frac{\nabla f}{f} = \frac{\mathbf{g}_0}{\alpha^2} = -\frac{\nabla W_0}{\alpha^2}, \tag{17}$$

where W_0 is the total gravity potential (gravitational plus centrifugal). Since the P-velocity may be regarded as constant on an equipotential surface, f is similarly seen to be constant on such surfaces and to be a function only of r_0, the mean radius of a given equipotential. It is then possible to write

$$\frac{df}{f} = -M^2 \frac{dr_0}{r_0}, \tag{18}$$

with a 'local compressibility number' defined as

$$M^2 = \frac{dW_0}{dr_0}\frac{r_0}{\alpha^2} = \frac{\bar{g}_0 r_0}{\alpha^2}, \tag{19}$$

where $\bar{g}_0$ is the mean gravity over the equipotential of radius r_0. Note that M^2 defined in this way is constant on an individual equipotential. If f_{0_i} is the value of f on an equipotential of mean radius r_{0_i}, integration gives $f(r_0)$ as

$$f(r_0) = f_{0_i} \exp - \int_{r_{0_i}}^{r_0} \frac{M^2}{r_0} dr_0. \tag{20}$$

A new functional is then formed by multiplying the left side of (13) through by

$$4\Omega^2\sigma^2\left(\sigma^2 - 1\right) f\chi^*$$

before integrating over the volume of the outer core, where σ a dimensionless angular frequency measured against twice the rotation angular frequency ($\sigma = \omega/2\Omega$). The new functional is then

$$\begin{aligned}\mathcal{F} &= 4\Omega^2\sigma^2\left(\sigma^2-1\right)\int \chi^* \nabla \cdot (f\mathbf{u})\, d\mathcal{V} \\ &= 4\Omega^2\sigma^2\left(\sigma^2-1\right)\left[\int \nabla \cdot (f\chi^*\mathbf{u})\, d\mathcal{V} \right. \\ &\left. - \int f\mathbf{u} \cdot \nabla\chi^* d\mathcal{V}\right].\end{aligned} \tag{21}$$

Before examining the properties of this new functional, it is useful to look at the form the equation of motion (1) takes under the subseismic approximation as expressed by (13). It may be written

$$\begin{aligned}&-\sigma^2\mathbf{u} + i\sigma\left(\hat{\mathbf{k}} \times \mathbf{u}\right) \\ &= -\frac{1}{4\Omega^2}\left[\nabla\chi - \frac{(1-\eta)}{\alpha^2}(\mathbf{u} \cdot \mathbf{g}_0)\mathbf{g}_0\right] = \boldsymbol{l},\end{aligned} \tag{22}$$

where $\hat{\mathbf{k}}$ is the unit vector in the direction of the uniform rotation. Thus, the components of the displacement field $\mathbf{u}$ and those of $\nabla\chi$ are linearly related and it is easily shown that

$$\mathbf{u} \cdot \mathbf{g}_0 = -\frac{\alpha^2}{1-\eta}\frac{\mathbf{C} \cdot \nabla\chi}{B}, \tag{23}$$

with

$$B = \frac{4\alpha^2\Omega^2}{1-\eta}\sigma^2\left(\sigma^2-1\right) + \sigma^2 g_0^2 - \left(\hat{\mathbf{k}} \cdot \mathbf{g}_0\right)^2, \tag{24}$$

$$\mathbf{C} = \left(\hat{\mathbf{k}} \cdot \mathbf{g}_0\right)\hat{\mathbf{k}} + i\sigma\hat{\mathbf{k}} \times \mathbf{g}_0 - \sigma^2\mathbf{g}_0. \tag{25}$$

It follows that

$$\boldsymbol{l} = -\frac{1}{4\Omega^2}\left[\nabla\chi + \mathbf{g}_0\frac{\mathbf{C}\cdot\nabla\chi}{B}\right]. \quad (26)$$

The general solution of (22) for $\mathbf{u}$ in terms of $\boldsymbol{l}$, valid for $\sigma^2 \neq 0$ or 1, is

$$\mathbf{u} = \frac{1}{\sigma^2(\sigma^2-1)}\left[\hat{\mathbf{k}}\left(\hat{\mathbf{k}}\cdot\boldsymbol{l}\right) - i\sigma\left(\hat{\mathbf{k}}\times\boldsymbol{l}\right) - \sigma^2\boldsymbol{l}\right]. \quad (27)$$

Substitution of expression (26) for $\boldsymbol{l}$ then produces

$$\mathbf{u} = \frac{1}{4\Omega^2\sigma^2(\sigma^2-1)}\left[\sigma^2 - \hat{\mathbf{k}}\hat{\mathbf{k}}\cdot - \frac{\mathbf{C}^*\mathbf{C}\cdot}{B} + i\sigma\hat{\mathbf{k}}\times\right]\nabla\chi. \quad (28)$$

It is therefore possible to regard χ as a generalized displacement potential and to write

$$\mathbf{u} = \mathcal{T}\nabla\chi, \quad (29)$$

where $\mathcal{T}$ is a second order Hermitean tensor with Cartesian components

$$T_{ij} = \frac{1}{4\Omega^2\sigma^2(\sigma^2-1)}\left[\sigma^2\delta_j^i - k_ik_j - \frac{C_i^*C_j}{B} + i\sigma\xi_{ilj}k_l\right]. \quad (30)$$

Here k_1, k_2, k_3 are the components of the unit vector $\hat{\mathbf{k}}$, δ_j^i is the Kronecker delta and ξ_{ilj} is the alternating tensor.

Both equations (13) and (16) can be reduced to an equation in χ alone, referred to as the 'subseismic equation' (Smylie et al, 1984). On substituting from (29) and (30) and scaling appropriately, they produce

$$\begin{aligned} &4\Omega^2\sigma^2\left(\sigma^2-1\right)\nabla\cdot(\mathcal{T}\nabla\chi) \\ &= \nabla\cdot\left[\sigma^2\nabla\chi - \hat{\mathbf{k}}\hat{\mathbf{k}}\cdot\nabla\chi - \mathbf{C}^*\frac{\mathbf{C}\cdot\nabla\chi}{B} + i\sigma\hat{\mathbf{k}}\times\nabla\chi\right] \\ &= \sigma^2\nabla^2\chi - \left(\hat{\mathbf{k}}\cdot\nabla\right)^2\chi - \mathbf{C}^*\cdot\nabla\left(\frac{\mathbf{C}\cdot\nabla\chi}{B}\right) \\ &-\frac{\mathbf{C}\cdot\nabla\chi}{B}\nabla\cdot\mathbf{C}^* = -4\Omega^2\sigma^2\left(\sigma^2-1\right)\frac{1}{\alpha^2}\mathbf{u}\cdot\mathbf{g}_0. \end{aligned} \quad (31)$$

Since

$$\nabla\cdot\mathbf{C}^* = A - \frac{4\Omega^2\sigma^2(\sigma^2-1)}{1-\eta}, \quad (32)$$

where

$$A = \frac{4\Omega^2\sigma^2(\sigma^2-1)}{1-\eta} + \sigma^2\left(4\pi G\rho_0 - 2\Omega^2\right) + \left(\hat{\mathbf{k}}\cdot\nabla\right)\hat{\mathbf{k}}\cdot\mathbf{g}_0, \quad (33)$$

on substitution from (23), we obtain the subseismic equation,

$$\sigma^2\nabla^2\chi - \frac{\partial^2\chi}{\partial z^2} - \frac{A}{B}\mathbf{C}\cdot\nabla\chi - \mathbf{C}^*\cdot\nabla\left(\frac{\mathbf{C}\cdot\nabla\chi}{B}\right) = 0. \quad (34)$$

With the application of the divergence theorem and substitution from (29) and (30), the functional given in (21) takes the form

$$\begin{aligned} \mathcal{F} = &-\sigma^2\int f\nabla\chi\cdot\nabla\chi^* d\mathcal{V} + \int f\frac{\partial\chi}{\partial z}\frac{\partial\chi^*}{\partial z}d\mathcal{V} \\ &+\int fB\psi\psi^* d\mathcal{V} - i\sigma\int f\hat{\mathbf{k}}\cdot(\nabla\chi\times\nabla\chi^*)\,d\mathcal{V} \\ &+4\Omega^2\sigma^2\left(\sigma^2-1\right)\int f\chi^*\mathbf{u}\cdot\hat{\mathbf{n}}dS, \end{aligned} \quad (35)$$

with

$$\psi = \frac{\mathbf{C}\cdot\nabla\chi}{B}. \quad (36)$$

Further treatment of the surface integral requires consideration of the boundary conditions.

Boundary Conditions and Deformation of the Elastic Shell

In the fluid outer core, under the subseismic approximation, both the displacement field and its conjugate are solenoidal after multiplication by f, according to equation (16). As well, owing to the Hermiticity of the tensor $\mathcal{T}$ appearing in equations (29) and (30), we have

$$\mathbf{u}\cdot\nabla\chi^* = \mathbf{u}^*\cdot\nabla\chi. \quad (37)$$

It then follows that the surface integral in expression (35) for the functional is real since

$$\begin{aligned} &\int f\chi^*\mathbf{u}\cdot\hat{\mathbf{n}}dS = \int\nabla\cdot(f\chi^*\mathbf{u})\,d\mathcal{V} \\ &= \int f\mathbf{u}\cdot\nabla\chi^* d\mathcal{V} = \int f\mathbf{u}^*\cdot\nabla\chi d\mathcal{V} \\ &= \int f\chi\mathbf{u}^*\cdot\hat{\mathbf{n}}dS. \end{aligned} \quad (38)$$

Thus, we need only consider functionals in which the surface integral can be replaced by

$$\frac{1}{2}\int f\left(\chi^*\mathbf{u} + \chi\mathbf{u}^*\right)\cdot\hat{\mathbf{n}}dS, \quad (39)$$

because at the solution points it is bound to be real.

In the solid parts of the Earth, the deformation is described by the sixth order spheroidal system of differential equations which govern the static field (Smylie and Mansinha, 1971) for spherical harmonic constituents of a particular degree n. For this linear system, the general solution can be written

$$\mathbf{y} = C_1\mathbf{y}_1 + C_2\mathbf{y}_2 + C_3\mathbf{y}_3 + C_4\mathbf{y}_4 + C_5\mathbf{y}_5 + C_6\mathbf{y}_6, \quad (40)$$

where the $\mathbf{y}$'s are six vectors with components representing the deformation field in the y-notation of Alterman, Jarosch and Pekeris (1959). The vectors $\mathbf{y}_1, \ldots, \mathbf{y}_6$ are the six fundamental solutions of the sixth order system. In the shell, for example, these are generated by setting $y_{ij}(b) = \delta_j^i$ with b the mean radius of the core-mantle boundary and y_{ij} the i-th component of $\mathbf{y}_j$. The integration is then carried forward to the surface at mean radius d.

At the free surface of the Earth, normal and shear stress vanish and the perturbation in gravitational potential becomes harmonic, while at the core-mantle boundary, shear stress vanishes and normal displacement, normal stress, gravitational potential perturbation and its gradient are all continuous. Application of these boundary conditions results in the linear algebraic system

$$\begin{aligned}
&C_1 y_{21}(d) + C_2 y_{22}(d) + C_3 y_{23}(d) \\
&+C_5 y_{25}(d) + C_6 y_{26}(d) = 0, \\
&C_1 y_{41}(d) + C_2 y_{42}(d) + C_3 y_{43}(d) \\
&+C_5 y_{45}(d) + C_6 y_{46}(d) = 0, \\
&C_1 z_1(d) + C_2 z_2(d) + C_3 z_3(d) \\
&+C_5 z_5(d) + C_6 z_6(d) = 0, \qquad (41) \\
&g_0(b) C_1 - \frac{1}{\rho_0(b^-)} C_2 - C_5 = \chi_n, \\
&4\pi G \rho_0 \left(b^-\right) C_1 + C_6 = \frac{dV_{1_n}}{dr_0}.
\end{aligned}$$

Here we have used the shorthand $z_i = y_{6i}(d)\, d/(n+1) + y_{5i}(d)$ and χ_n is the degree n radial coefficient of the spherical harmonic expansion of χ on the core-mantle boundary, dV_{1_n}/dr_0 the derivative of the degree n radial coefficient of the spherical harmonic expansion of V_1 on the outer core surface with respect to mean radius.

For $n = 0$, the sixth order spheroidal system degenerates to fourth order and the algebraic system (41) also reduces to fourth order.

It is therefore possible to write

$$y_1(b) = \frac{1}{g_0(b)} \left[h_n^1 \chi_n(b) - b h_n^2 \frac{dV_{1_n}}{dr_0}(b) \right], \qquad (42)$$

giving the degree n part of the normal displacement at the core-mantle boundary in terms of Love-like coefficients h_n^1, h_n^2 which can be calculated by solving the algebraic system (41). The appropriate coefficients found for the shell, using Earth model 1066A of Gilbert and Dziewonski (1975), are listed in Table 1.

The surface integral (39) appearing in the functional, by virtue of the orthogonality relation among Legendre functions, is then expressible as

$$\frac{4\pi b^2}{g_0(b)} \sum_{n=m} \frac{1}{2n+1} \frac{(n+m)!}{(n-m)!} \left[h_n^1 \chi_n \chi_n^* - b h_n^2 \frac{d}{dr_0} \left(\frac{V_{1_n} V_{1_n}^*}{2} \right) \right] \qquad (43)$$

for modes of azimuthal number m, the common factor f_{0_i} appearing in relation (20) having been set to unity at $r_{0_i} = b$.

TABLE 1. Internal Load Love Numbers for the Shell Computed for Earth Model 1066A of Gilbert and Dziewonski (1975).

Degree n	Load number h_n^1	Load number h_n^2
0	0.6076	0.3800
1	-15.8014	-7.9588
2	0.6159	0.1251
3	0.3623	0.0501
4	0.2381	0.0262
5	0.1750	0.0163
6	0.1398	0.0113
7	0.1177	0.0084
8	0.1025	0.0066
9	0.0911	0.0053
10	0.0822	0.0043
11	0.0749	0.0036
12	0.0689	0.0031
13	0.0638	0.0027
14	0.0594	0.0023
15	0.0556	0.0020
16	0.0523	0.0018
17	0.0493	0.0016
18	0.0467	0.0014
19	0.0443	0.0013
20	0.0422	0.0012
21	0.0402	0.0011
22	0.0385	0.0010
23	0.0369	0.0009

Numerical Implementation

The functional to be utilized in numerical calculations can now be written

$$\begin{aligned}
\mathcal{F} = &-4\Omega^2\sigma^2\left(\sigma^2 - 1\right) \int f \mathbf{u} \cdot \nabla \chi^* d\mathcal{V} \\
&+2\Omega^2\sigma^2\left(\sigma^2 - 1\right) \int f\left(\chi^* \mathbf{u} + \chi \mathbf{u}^*\right) \cdot \hat{\mathbf{n}} dS, \qquad (44)
\end{aligned}$$

the surface integral being given by (43) and the volume integral expanded as in expression (35).

Under the subseismic approximation (13) in the core, the system (41) becomes degenerate since then

$$g_0(b) C_1 - \frac{1}{\rho_0(b^-)} C_2 = 0. \qquad (45)$$

Each harmonic of χ is subject to the mixed boundary condition

$$\chi_n - \frac{b_n}{(n+1)} \frac{dV_{1_n}}{dr_0} = 0. \qquad (46)$$

Together with the identity (37) derived from the Hermiticity of the tensor $\mathcal{T}$, this ensures that the variation of the functional (44) can be written

$$\delta\mathcal{F} = 4\Omega^2\sigma^2\left(\sigma^2-1\right)\int[\delta\chi\nabla\cdot(f\mathbf{u}^*)$$
$$+\delta\chi^*\nabla\cdot(f\mathbf{u})]\,d\mathcal{V}$$
$$-4\Omega^2\sigma^2\left(\sigma^2-1\right)\int f\left[\delta\chi\left(\mathbf{u}^*-\mathbf{u}_b^*\right)\cdot\hat{\mathbf{n}}\right.$$
$$\left.+\delta\chi^*\left(\mathbf{u}-\mathbf{u}_b\right)\cdot\hat{\mathbf{n}}\right]dS, \tag{47}$$

where in accord with expression (42),

$$\int\left[\delta\chi\mathbf{u}_b^*\cdot\hat{\mathbf{n}}+\delta\chi^*\mathbf{u}_b\cdot\hat{\mathbf{n}}\right]dS$$
$$=4\pi b^2\sum_{n=m}\frac{1}{2n+1}\frac{(n+m)!}{(n-m)!}\left[\delta\chi_n y_1^*(b)\right.$$
$$\left.+\delta\chi_n^* y_1(b)\right]. \tag{48}$$

For variations in χ which satisfy (46), stationarity of $\mathcal{F}$ is then seen to imply not only satisfaction of the subseismic equation (16) but also the boundary conditions expressed by the system (41). Thus, we have suceeded in constructing a functional which obeys the 'natural' boundary conditions of the problem.

Once the problem has been solved for the scalar χ and the displacement field $\mathbf{u}$, the Poisson equation (2) can be solved for the gravitational potential perturbation V_1 with the density perturbation computed from (6) and (13) as

$$\rho_1 = \frac{\rho_0}{\alpha^2}(1-\eta)\,\mathbf{u}\cdot\mathbf{g}_0. \tag{49}$$

Equation (46) then provides the Neumann boundary condition on V_1 required to make the solution unique.

The calculation is implemented numerically via a representation of the dependent variable χ in piecewise bicubic, Hermite splines. The integrands in the functional, and in the expansion of the boundary values of χ in spherical harmonics, are arranged to have polynomial forms by suitable transformation of the dependent variable so that they may be calculated exactly by Gaussian integration. The functional (44) then becomes a symmetric bilinear form whose stationarity is ensured by satisfaction of the corresponding linear, homogeneous algebraic system. Details of this implementation and of the technique used to ensure that condition (46) is satisfied will be presented in a full paper to be published elsewhere.

Discussion

The dynamical behaviour of the fluid outer core at periods which are a substantial fraction of the rotation period and longer has much in common with classical theory and modern laboratory experiments but it also differs in several essential ways due to the realistic properties of the Earth. While compressibility effects can be handled by a relatively simple 'decompression' factor f in the functional, stratification gives rise to a whole new branch of core physics that awaits investigation. Analysis of the conditions at the elastic boundaries of the fluid core also reveals a significant departure from the usual rigid assumption of laboratory models. Because of the planetary scale of the Earth, pressure variations arising from transport through the gravitationally induced static pressure field dominate by nearly three orders of magnitude. In turn, the condition to be applied at the boundaries more closely resembles a potential condition than the usual rigid requirement.

Acknowledgments. The author is grateful to Dr. A.M.K. Szeto for discussions of many technical points.

References

Aldridge, K. D., Axisymmetric inertial oscillations of a fluid in a rotating spherical shell, *Mathematika,* **19**, 163-168, 1972.

Aldridge, K. D. and Lumb, L. I., Inertial waves identified in the Earth's fluid outer core, *Nature,* **325**, 421-423, 1987.

Aldridge, K. D. and Toomre, A., Axisymmetric inertial oscillations of a fluid in a rotating spherical container, *J. Fluid Mech.,* **37**, 307-323.

Bryan, G. H., The waves on a rotating liquid spheroid of finite ellipticity, *Phil. Trans. R. Soc. Lond.* A, **180**, 187-219, 1889.

Friedlander, S., Internal oscillations in the Earth's fluid core, *Geophys. J. R. astr. Soc.*, **80**, 345-361, 1985.

Gilbert, F. and Dziewonski, A.M., An application of normal mode theory to the retrieval of structural parameters and source mechanisms from seismic spectra, *Phil. Trans. R. Soc. Lond.* A, **278**, 187-269, 1975.

Greenspan, H.P., *The Theory of Rotating Fluids,* Cambridge University Press, Cambridge, 1969.

Johnson, I. M. and Smylie, D. E., A variational approach to whole-Earth dynamics, *Geophys. J. R. astr. Soc.,* **50**, 35-54, 1977.

Melchior, P. and Ducarme, B., Detection of inertial gravity oscillations in the Earth's core with a superconducting gravimeter at Brussels, *Phys. Earth planet. Interiors,* **42**, 129-134, 1986.

Poincaré, H., Sur l'équilibre d'une masse fluide animée d'un mouvement de rotation, *Acta Math.,* **7**, 259-380, 1885.

Smylie, D.E., Variational calculation of wobble modes of the Earth, (this issue) 1988.

Smylie, D. E. and Mansinha, L., The elasticity theory of dislocations in real Earth models and changes in the rotation of the Earth, *Geophys. J. R. astr. Soc.,* **23**, 329-354, 1971.

Smylie, D. E. and Szeto, A. M. K., The dynamics of the Earth's inner and outer cores, *Rep. Prog. Phys.,* **47**, 855-906, 1984.

Stewartson, K. and Rickard, J. A., Pathological oscillations of a rotating fluid, *J. Fluid Mech.,* **35**, 759-773, 1969.

CORE UNDERTONES IN AN ELLIPTICAL UNIFORMLY ROTATING EARTH

Véronique Dehant[1]

Institut d'Astronomie et de Géophysique G. Lemaître, Université Catholique de Louvain, 2, Chemin du Cyclotron, B 1348 Louvain-la-Neuve, Belgique

Abstract. The response of the Earth to an external force is perturbed by resonance effects. The core normal modes are usually ignored because they are considered to be confined in the liquid core. Recent gravity data analysis [Melchior and Ducarme, 1986] has revealed some unexplained periods around 14h that might be related to core normal modes. Various works prepared by "core equation specialists" has been reviewed in order to introduce the core mode effects in the last adopted Earth model i.e. an uniformly rotating, elliptical Earth with an elastic inner core, a liquid core and an elastic mantle. An attempt is made to account for core modes in an elliptical, rotating, compressible and inviscid fluid outer core.

Introduction

The aim of this work is to concentrate our effort on the theory of core dynamics. The liquid outer core has been recognized responsible for features and processes observed at the surface of the Earth (like the non dipole magnetic field drift). The last adopted model for the computation of tides and nutations [Smith, 1974 and Wahr, 1979], is an elliptical, uniformly rotating Earth with an elastic inner core, a liquid outer core and an elastic mantle. In this model, the equations inside the core are rather simplified. The question which arises at this stage is: *What is the contribution of the real core to the tides and the nutations?* It is necessary to make this point because a difference exists nowadays between the observations and the theory [Herring et al., 1986] and it is still not explained. Gwinn et al. [1986] pointed it out for the nutations. For the tides, Melchior and De Becker [1983] computed a discrepancy of about 1.5 percent in the tidal gravimetric factors δ. This difference is reduced by a factor of two if the same definition of δ is used for the theory as for the observations [Dehant and Ducarme, 1987]. Still, there remains a gap of about 0.7 percent.

The tidal observations with the superconducting gravimeter reveal some peaks close to, but out of the semi-diurnal band, as pointed out by Melchior and Ducarme [1986] and reconfirmed later for a 4 years data set analysis [Melchior et al., 1987]. Melchior and Ducarme [1986] suggest that these peaks may reflect the excitation of inertial waves in the core by earthquakes or that they may be related to translational modes of the inner core. This question is still open and the aim of our work would be to answer the question: *Are these peaks related to core modes? Are they inertial-gravity waves or gravity-inertial waves?*

These two reasons convince us that it is necessary to use much more complete equations in the liquid core. This extension concerns the relative magnitude of the Coriolis and buoyancy forces. Core dynamics is then reviewed in the first part of this paper and in the second part, an attempt is presented to integrate, in the liquid outer core part of the Smith and Wahr's model, the just mentioned more complete equations.

Review of Core Dynamics

General Formulation of the Equations

Before establishing the equations, let us clarify the hypotheses that are sustained in this work:
1. the Earth is hydrostatically pre-stressed,
2. it is uniformly rotating at an angular speed Ω,

[1] **Senior Research Assistant - National Fund for Scientific Research (Belgium)**

3. it is ellipsoidal,
4. the liquid core is stratified,
5. the frequency band we are working in is the tidal band and accept that no viscosity and no magnetic field are considered [Crossley and Smylie, 1975].

The response of the Earth to the external tidal force can be expressed by a set of vectorial equations:

1. The vectorial equation of motion:

$$-\rho_0\omega^2\vec{u} + 2i\omega\rho_0\vec{\Omega} \wedge \vec{u} = \\ -\vec{\nabla}p^E - \rho_1\vec{\nabla}V_0 - \rho_0\vec{\nabla}V_1 - \rho_0\vec{\nabla}W^T$$

where ρ_0 is the density at the pre-equilibrium state, ρ_1 is the eulerian increment of density, $\vec{u}$ is the displacement field, p^E is the eulerian pressure, V_0 is the initial gravity potential, V_1 is the eulerian potential (due to mass redistribution associated to the deformations), W^T is the external tidal potential.

2. The constitutive equation (the stress-strain relationship):

$$p^E = -\vec{u}.\vec{\nabla}p_0 - k\vec{\nabla}.\vec{u} \tag{1}$$

$$p^L = -k\vec{\nabla}.\vec{u} \tag{2}$$

where p^L is the lagrangian pressure, p_0 is the hydrostatic equilibrium pressure and k is the bulk modulus.

3. Poisson's equation:

$$\nabla^2V_1 = 4\pi G\rho_1 \tag{3}$$

Usually, a reduced pressure is introduced:

$$\phi = \frac{p^E}{\rho_0} + V_1(+W^T), \tag{4}$$

so that (1) becomes:

$$-\rho_0\omega^2\vec{u} + 2i\omega\rho_0\vec{\Omega} \wedge \vec{u} + \rho_0\vec{\nabla}\phi - \frac{N^2}{g_0}p^L\hat{e}_r = 0 \tag{5}$$

where N^2 is the square of Brunt-Väisälä frequency and g_0 is the initial gravity.

In order to transform these vectorial equations into scalar equations, i.e. the Partial Differential Equations (PDE) into Ordinary Differential Equations (ODE) of the first order in $\frac{d}{dr}$, it is necessary to use expansions involving either the Generalized Spherical Harmonic functions (GSH) of Phinney and Burridge [1973] or the Associate Legendre Polynomials (ALP).

In the non rotating case, one system of 4 equations with 4 spheroidal variables (σ_l^m) inside the liquid core is obtained for each degree and order (l, m) of the external potential in the case of forced motion, and of the GSH in the case of free motion. Let A_0 be the operator in this case. Now, if the Earth is rotating, there is a coupling between the spheroidal displacement of the order l (σ_l^m) and the toroidal displacements of the order $l+1$ (τ_{l+1}^m) and $l-1$ (τ_{l-1}^m) in the system. Moreover, if the Earth's ellipticity is introduced, the spheroidal variables of the order l are coupled with the toroidal variables of the order $l+1$ and $l-1$ and with the spheroidal variables of the order $l+2$ (σ_{l+2}^m) and $l-2$ (σ_{l-2}^m); so that finally, an infinite number of equations and of variables appears necessarilly in each system (see Figure 1).

In the theory of Smith and Wahr, each system is truncated as presented in Figure 2. In both figures, the notation a is used for an operator proportional to Ω and b, for an operator proportional to the ellipticity ϵ. Truncation is allowed in the inner core and the mantle, but in the outer core, the Coriolis force is much more important and this is the reason why no truncation is allowed at the first order in ϵ or Ω and moreover, it will be necessary to expand the series very far.

In the literature, two approximations can be found, which reduce the number of equations, or avoid a spherical harmonics expansion. They are respectively the Boussinesq approximation and the Subseismic approximation.

The Boussinesq Approximation

The Boussinesq approximation has been introduced by Crossley and Rochester in 1980 and used by Crossley [1984] for numerical computation in the case of a spherical Earth. Their hypothesis is: *The variation of the density is neglected everywhere except in the buoyancy terms.* This approximation implies the incompressibility condition inside the core. The reduced pressure (4) does not contain V_1 any more in this case, so that the motion and constitutive equations are decoupled from Poisson's equation. The number of equations is reduced by a factor 2. In their approximation, the Brunt-Väisälä frequency, N, reduces to the following form:

$$N^2 = -\frac{g_0}{\rho_0}\frac{d\rho_0}{dr} \tag{6}$$

This constrains Crossley and Rochester [1980] to redefine N^2 and to consider that: *The Brunt-Väisälä frequency is a parameter computed from seismic data.* They use then the normal definition of N^2 ignoring their first approximation at this step.

The Subseismic Approximation

The name of this approximation is not exactly related to the approximation itself, but related to the general hypothesis 5 (see paragraph 2.1). It expresses that the fre-

$$
\begin{pmatrix}
 & & & & & & & & \\
a+b & a+b & A_0+a+b & a+b & a+b & & & & \\
 & a+b & a+b & A_0+a+b & a+b & a+b & & & \\
 & & a+b & a+b & A_0+a+b & a+b & a+b & & \\
 & & & a+b & a+b & A_0+a+b & a+b & a+b & \\
 & & & & a+b & a+b & A_0+a+b & a+b & a+b \\
 & & & & & & & &
\end{pmatrix}
$$

$$
\begin{pmatrix} \cdot \\ \cdot \\ \cdot \\ \sigma_{l-2}^m \\ \tau_{l-1}^m \\ \sigma_l^m \\ \tau_{l+1}^m \\ \sigma_{l+2}^m \\ \cdot \\ \cdot \\ \cdot \end{pmatrix} = \begin{pmatrix} \cdot \\ \cdot \\ \cdot \\ 0 \\ 0 \\ 0 \\ 0 \\ 0 \\ \cdot \\ \cdot \\ \cdot \end{pmatrix}
$$

Fig. 1. Infinite number of equations for one system in the liquid core.

quencies we are working with are lower than the seismic frequencies but not too far, i.e. subseismic frequencies.

The Subseismic approximation has been introduced by Smylie and Rochester in 1981. They consider first the general hypothesis of subseismic frequencies and they add the following consideration: *The dynamic pressure can be negligibly small in the determination of the density variation.* Then, the eulerian density can be related to the initial gravity and the Brunt-Väisälä frequency by:

$$\rho_1 = \frac{\rho_0}{g_0} N^2(\vec{u}.\hat{e}_r) \tag{7}$$

In this case, the cubic dilatation ($\vec{\nabla}.\vec{u}$) is not zero.

After some manipulations of the whole set of vectorial equations and after introduction of new variables, they set up a vectorial equation called the Subseismic Wave Equation (SWE) (this may be the reason of the name of the approximation). Then, using the variational principle, they integrate a set of 3 scalar equations at 3 variables, including the SWE, and from there, they can compute ϕ, $\vec{u}.\hat{e}_r$ and $\vec{\nabla}.\vec{u}$. From these results, it is now possible to deduce ρ_1, and using ρ_1, to solve for V_1 (for more details, see [Smylie and Rochester, 1981, 1985 and 1986]).

Comparison Between Both Approximations

The equations of motion are exactly identical in both approximations, although the physical meaning of the variables are different. For example, the reduced pressure ϕ is defined in a different way (see Table 1). In the Boussinesq approximation, the constitutive equation is replaced by the incompressibility condition. But Fried-

$$
\begin{pmatrix}
A_0 & a+b & a+b & 0 & 0 \\
0 & A_0+a+b & a+b & a+b & 0 \\
0 & a+b & A_0+a+b & a+b & 0 \\
0 & a+b & a+b & A_0+a+b & 0 \\
0 & 0 & a+b & a+b & A_0
\end{pmatrix}
\begin{pmatrix} \sigma_{l-2}^m \\ \tau_{l-1}^m \\ \sigma_l^m \\ \tau_{l+1}^m \\ \sigma_{l+2}^m \end{pmatrix} = \begin{pmatrix} 0 \\ 0 \\ 0 \\ 0 \\ 0 \end{pmatrix}
$$

Fig. 2. Truncated system in the theory of Smith and Wahr.

TABLE 1.

	Boussinesq approximation Crossley & Rochester (1980)	Subseismic approximation Smylie & Rochester (1981)
ϕ	$\simeq \frac{p^E}{\rho_0}$	$= \frac{p^E}{\rho_0} + V_1$
Compressibility	no: $\vec{\nabla}.\vec{u} = 0$	yes: $\vec{\nabla}.\vec{u} = \frac{\rho_0 g_0}{k}(\vec{u}.\hat{e}_r)$
N^2	from seismic profile non classical definition $N^2 = -\frac{g_0}{\rho_0}\frac{d\rho_0}{dr}$	from seismic profile classical definition $N^2 = -\frac{g_0}{\rho_0}\frac{d\rho_0}{dr} - \frac{\rho_0 g_0^2}{k}$
ρ_1	$\rho_1 = -\frac{d\rho_0}{dr}(\vec{u}.\hat{e}_r)$	$\rho_1 = \frac{\rho_0}{g_0}N^2(\vec{u}.\hat{e}_r)$
p^L	$p^L = -\lim_{\substack{k \to \infty \\ \vec{\nabla}.\vec{u} \to 0}} k\vec{\nabla}.\vec{u}$ p^L is a variable	$p^L = -k\vec{\nabla}.\vec{u}$
Equation of motion Poisson's equation Constitutive equation	form (17) decoupled from (17) $\vec{\nabla}.\vec{u} = 0$	form (17) coupled: $\nabla^2 V_1 = 4\pi G\rho_1$ $p^L = k\vec{\nabla}.\vec{u}$

lander [1986] shows that the expression of the dilatation in the Subseismic approximation, can be written:

$$\vec{\nabla}.(\rho_0^\star \vec{u}) = 0 \qquad (8)$$

where

$$\rho_0^\star = e^{\int \frac{g_0^2}{\rho_0 g_0^2 + N^2 k}\frac{d\rho_0}{dr}dr} \qquad (9)$$

$\rho_0^\star$ then plays the role of the Boussinesq approximation density. It may be called "density" but it has not the same physical meaning as the true density. The equations using that formulation have exactly the similar mathematical formulation although the "density" and the reduced pressure have two different physical meanings.

Friedlander [1985] points out that the presence of the compressibility factor leaves the fundamental nature of the spectrum unchanged. She emphasizes that it is unchanged in terms of stability, discreteness, limit points and bounds, although each eigenvalue is shifted slightly by a term dependent on the magnitude of $\frac{g_0\rho_0}{k}$ [Friedlander, 1987].

Elliptical, Uniformly Rotating Earth

From the previous review, one concludes that in the core, two important effects, giving two kinds of symmetry in the problem, are found:

1. the Coriolis force, to which a cylindrical symmetry is associated,
2. the buoyancy force, related to the stratification effect, to which a spherical symmetry is associated.

As already mentioned, the scalar equations in the mantle and in the inner core are expressed using either the GSH or the ALP. The system of equations is truncated. In the liquid core, due to the same relative importance of the magnitudes of the two forces cited above, it is not allowed to truncate: there is an infinite coupling between all the variables. Although, truncation is presently used in the last adopted model (Smith and Wahr model). The aim of our work is to find a way to solve this problem. In the scope of the present paper, we consider two possibilities:

1. to apply Crossley and Rochester's method using GSH expansion and truncate; but, it will be necessary to expand the series very far;

TABLE 2.

1st	step :	4 equations
2d	step :	18 equations
3st	step :	30 equations
4st	step :	42 equations
5st	step :	54 equations
6st	step :	64 equations

2. to apply Smylie and Rochester's method, solving the Subseismic Wave Equation for the new variables; in this case, because GSH are used in the mantle, new boundary conditions must be found, linking the new variables and the classical variables introduced in the mantle. Again, this implies an infinite series expansion and again, truncation must necessarilly happen very far; but in this case, the equations to solve are no longer PDE but algebric equations; it would then probably be much easier to take more terms in the series expansion.

As in the second case it is necessary to combine two computer programs (the Subseismic Equation solution for the core and Wahr's solution for the mantle), we decided, for this paper, to limit our tests to the GSH expansion (first possibility) and compute step by step the new coefficients in the series. Unfortunately, we could only write down the equations till step number three because of the complexity of the operators and because of the important increase of the number of equations. As an example, 64 ODE are to be solved at the step number 6! (see Table 2)

We then decide to investigate the sensitivity of the computation to Earth rotation. We know that, if there is no rotation, truncation is allowed in the core. We took a very small rotation and computed some core modes using Smith and Wahr's programs. By changing the rotation a little bit, new normal modes could be found. They very rapidly shift so that one can conclude that numerical instabilities are very important in this computation.

Conclusion

Due to the non convergence of the numerical computation, it is not possible to calculate the core modes using GSH expansion. This confirms what has already been pointed out in the case of a spherical Earth by Crossley and Rochester [1982]. They try to identify the eigenperiods and pointed out the increase of the uncertainty when increasing the order of the spherical harmonics representation. In order to answer the questions presented in the introduction, we intend first to try to find analytical solutions at the CMb and to propagate them in the mantle. For that part of our future work, we shall work with Kudlick's solution [1966]. Secondly, we shall try to use finite element integration inside the core in order to find new numerical solutions.

Acknowledgments. The core dynamics review presented in this paper was possible thanks to discussions with D.J. Crossley, S. Friedlander, M.G. Rochester and D.E. Smylie (in alphabetic order). We are thankful to them for all the advices they gave us. We wish to dedicate our work to them and to Prof. P. Melchior who initiates all our investigations in this field by writing his book in the Physics of the Earth's Core [1986].

References

Crossley, D.J., Oscillatory flow in the liquid core., Phys. Earth Planet. Inter., 36, 1-16, 1984.

Crossley, D.J. and Rochester, M.G., Simple core undertones., Geophys. J. R. astr. Soc., 60, 129-161, 1980.

Crossley, D.J. and Smylie, D.E., Electromagnetic and viscous damping of core oscillations., Geophys. J. R. astr. Soc., 42, 1011-1033, 1975.

Dehant, V. and Ducarme, B., Comparison between the observed and the computed tidal gravimetric factor., Phys. Earth Planet. Inter., 49, 192-212, 1987.

Friedlander, S., Stability of the subseismic wave equation for the Earth's fluid core., Geophys. Astrophys. Fluid Dynamics, 31, 151-167, 1985.

Friedlander, S., Stability and waves in the Earth's fluid core., Proceedings of US-Italy conference on Energy Stability and Convection, Capri, May 1986, 21 pp, 1986.

Friedlander, S., Internal waves in a rotating stratified spherical shell: asymptotic solutions., Geophys. J. R. astr. Soc., 89, 637-655, 1987.

Gwinn, C.R., Herring, T.A. and Shapiro I.I., Geodesy by Radio Interferometry: studies of the forced nutations of the Earth. 2. Interpretation., J. Geophys. Res., 91, b5, 4755-4765, 1986.

Herring, T.A., Gwinn, C.R. and Shapiro I.I., Geodesy by Radio Interferometry: studies of the forced nutations of the Earth. 1. Data analysis., J. Geophys. Res., 91, b5, 4745-4754, 1986.

Kudlick, M.D., On transient motions in a contained, rotating fluid., Ph. D. thesis M.I.T., USA, 154 pp, 1966.

Melchior, P., The Physics of the Earth's Core. An Introduction., Pergamon Press, Great Britain, 256 pp, 1986.

Melchior, P., Crossley, D.J., Dehant, V. and Ducarme, B., Have inertial waves been identified from the Earth's core?, Proceedings of the Symposium U2 of the 19th General Assembly of the IUGG, Vancouver, Canada, 1987.

Melchior, P. and De Becker, M., A discussion of world-wide measurements of tidal gravity with respect to oceanic interaction, lithosphere heterogeneities, Earth's flattening and inertial forces., Phys. Earth Planet. Inter., 31, 27-53, 1983.

Melchior, P. and Ducarme, B., Detection of inertial gravity oscillations in the Earth's core with a superconducting gravimeter at Brussels., Phys. Earth Planet. Inter., Letter Section, 42, 129-134, 1986.

Phinney, R.A. and Burridge, R., Representation of the elastic-gravitational excitation of a spherical Earth model by Generalized Spherical Harmonics., Geophys. J. R. astr. Soc., 34, 451-487, 1973.

Smith, M.L., The scalar equations of infinitesimal elastic-gravitational motion for a rotating, slightly elliptical Earth., Geophys. J. R. astr. Soc., 37, 491-596, 1974.

Smylie, D.E. and Rochester, M.G., Compressibility, core dynamics and the subseismic wave equation. Phys. Earth Planet. Inter., 24, 308-319, 1981.

Smylie, D.E. and Rochester, M.G., Long period core dynamics., Proceedings of the NATO AWK workshop on Earth Rotation: Solved and Unsolved Problems., Bonas, France, ed. A. Cazenave, published in 1986, 297-324, 1985.

Smylie, D.E. and Rochester, M.G., A variational principle for the subseismic wave equation., Geophys. J. R. astr. Soc., 86, 553-561, 1986.

Wahr, J.M., The tidal motions of a rotating, elliptical, elastic and oceanless Earth., Ph. D. thesis, University of Colorado, 216 pp, 1979.

AN EXPERIMENTAL STUDY OF INERTIAL WAVES IN A SPHEROIDAL SHELL OF ROTATING FLUID

L. Ian Lumb and Keith D. Aldridge

Department of Earth and Atmospheric Science, York University,
North York, Ontario M3J 1P3, Canada

Abstract. Travelling inertial waves contained within a spheroidal shell of a homogeneous, uniformly rotating fluid have been studied experimentally. Both retrograde and prograde waves of azimuthal number unity have been identified. Significant interaction among modes, during the free decay of an excited mode, is readily observable from measured free decay rates. Applications of the inertial wave studies in this geometry are found in the dynamics of the Earth's fluid outer core.

Introduction

The importance of studies on inertial waves in contained rotating fluids to the dynamics of the Earth's fluid outer core has been shown by Aldridge and Lumb [1987] (hereinafter AL) and more recently by Aldridge et al. [this issue] (hereinafter ALH). AL interpreted the long-period gravimetric (LPG) data of Melchior and Ducarme [1986] in terms of the Poincaré Earth model. It was their expectation that departures from this model would describe the fluid properties of the real Earth. Since the Poincaré problem is a limiting case of a more realistic Earth model, such as Smylie and Rochester's [1981] *subseismic* equation, there exists the possibility of 'tracking' the results of the inertial-wave studies to the real Earth [Smylie and Rochester, 1986].

Inertial waves are important in other areas of geodynamics. Olson's [1981] *inertial wave dynamo*, for example, suggests the importance of these waves in ultimately understanding the generation and maintenance of the terrestrial magnetic field. In contrast, Szeto and Smylie [1984] proposed an inclined inner core model in an effort to explain the secular variation and reversals of the geomagnetic field. Other examples relate to the rotational dynamics of the Earth. Smylie [this issue] has recently shown how core modes, coupled through the mantle, yield an infinite suite of wobble modes. As a further example, Dehant [this issue] is concerned with inclusion of core modes to help explain the discrepancy between the theoretical and observed period of the free core nutation. Experimental verification of inertial waves and their properties provides much needed physical evidence to complement these theoretical studies.

The present series of experiments considers the motions of a uniformly rotating, incompressible, homogeneous fluid, contained within a spheroidal shell. The linear, inviscid response to perturbations is described by the Poincaré equation plus boundary conditions [e.g. Greenspan, 1968]. Closed form solutions have only been found for a fluid which fills a spherical cavity [Greenspan, 1968]. The inclusion of an inner boundary, however, causes the problem to be *ill-posed* and brings the existence of continuous solutions into question [Stewartson and Rickard, 1969]. This ill-posedness lead Aldridge [1972] to the development of a variational principle for the axisymmetric modes of the Poincaré problem. With this approximate method, Aldridge was able to partially interpret measured eigenfrequencies for modes in the spherical shell geometry. The extension of the variational principle to allow for non-axisymmetric modes has been made recently by Smylie and Rochester [1986].

A number of experiments [e.g. Aldridge and Toomre, 1969; Aldridge, 1972] have been carried out in the spherical geometry. A review of these investigations was given by Stergiopoulos and Aldridge [1984] (hereinafter SA). SA verified the *existence* of retrograde inertial waves in the spheroidal shell geometry experimentally. Their findings included the observation of a significant interaction between modes during the *free decay* of an excited mode.

The experiments described here differ from those of SA in the following ways: the tilt of the inner spheroid is altered to examine the effect of this nonlinearity on the interaction between modes;the existence of *prograde* inertial modes is demonstrated experimentally and finally, steps were taken to improve the resolution of the data.

Experimentally observed properties of inertial waves provide guidance for theoretical studies on their real Earth counterparts. In particular, effects like differential rotation, resonance collapse [e.g. Gunn and Aldridge, this issue], modal interactions, dissipation [e.g. AL] and boundary irregularities [e.g. ALH] are all important yet not considered by most theoretical treatments. It is our expectation that the inclusion of some of these experimentally observed effects will ultimately define a realistic model of the Earth's core.

Experimental Setup

Inertial waves of azimuthal wavenumber unity were excited by precessing a spheroid within a spherical cavity (see Figure 6 of SA). The inner spheroid is precessed at angular frequency ω while the container is rotated at speed Ω; the ratio of precession to rotation rate is the dimensionless eigenfrequency $\lambda^{\mathcal{R}}$ (i.e., the real part of the complex eigenfrequency $\lambda = \lambda^{\mathcal{R}} + i\,\lambda^{\mathcal{I}}$). The major axis of the inner spheroid is inclined to the equatorial plane of the container at an angle ϵ. The radius of the spherical cavity is 10.0 cm.

Inertial waves are travelling disturbances in the rotating frame. Pressure changes associated with their passage are measured via two ports which are connected to a differential pressure transducer. With the current arrangement, pressure changes on the order of a fraction of a millimeter of fluid are measurable.

The precessing spheroid and rotating container are controlled independently. Therefore, it is possible to excite both retrograde (precession and rotation in *opposite* senses) and prograde (precession and rotation in the *same* sense) modes. To excite a particular mode, the ratio ω/Ω is adjusted to 'tune in' to a specific resonance location. For example, to excite the fundamental or 'spin-over' mode, the inner spheroid is precessed in the retrograde sense at *precisely* the speed of rotation. Therefore, to an observer in the laboratory (i.e., the inertial frame), the inner spheroid appears stationary.

The chamber containing the fluid (silicone oil with nominal viscosity $\nu = 6.3$ cS) is mounted on a turntable (see Figure 7 of SA). As excitation of a particular inertial mode depends strongly on the ratio ω/Ω, it is important that these speeds are maintained to within a few tenths of a percent. This is achieved by independent servo-controllers for both the spheroid and turntable.

The servo-controllers and data acquisition are controlled by a Perkin-Elmer 5/16 minicomputer as depicted in Figure 7 of SA. Instructions for the servo-controllers are input to the experiment via the D/A converter while disturbance pressure signals are received by the A/D converter. The spheroid was precessed at frequency near 1 Hz while sampling of pressure was made at approximately 20 kHz. Electrical contact between the rotating system and the laboratory is maintained by two sets of slip rings located at the top and bottom of the turntable. Data collected during an experimental run is stored on floppy disc, while subsequent data analysis and processing are performed on a VAX 8600.

Greenspan [1968] has shown that the eigenvalues for the Poincaré equation are real-valued and bounded i.e., $\lambda^{\mathcal{R}} \leq |2|$. Examination of the appropriate eigenfunctions suggests that waves for which $\lambda^{\mathcal{R}} \approx 0$ have motions predominantly *parallel* to the axis of rotation; whereas, waves for which $\lambda^{\mathcal{R}} \approx 2$ have motions *perpendicular* to the rotational axis. Thus, the nature of the perturbing spheroid allows for *direct* excitation of modes having small $\lambda^{\mathcal{R}}$ and *indirectly* excites modes for large $\lambda^{\mathcal{R}}$. Therefore, during free decay, it is possible to indirectly excite the modes AL identified in the LPG data of Melchior and Ducarme [1986].

The experimental procedure is to precess the inner spheroid near one of its resonant points while the fluid rotates as a solid body. The resonance is well established after several *spin-up times,* at which point, the perturbing inner spheroid is switched off and the excited mode is allowed to freely decay. (For a fluid of viscosity ν and length scale R, the spin-up time, $T = R/(\nu\Omega)^{1/2}$, measures the time scale for the fluid to achieve solid body rotation [Greenspan, 1968] as well as the time scale for the free decay of inertial waves.) Measurement of the pressure field, as sensed by the differential pressure transducer, commences just prior to switch off and continues for $2-3$ spin-up times.

Data Processing

The data recorded from an experimental run are in the form of a time series for the measured disturbance pressure. It is evident from observation that the form of this series is that of multiple exponential decays, so that the data are modelled to be M simultaneously decaying sinusoids. For N data points, the calculated response may be written

$$P_i = \sum_{j=1}^{M} A_j e^{-\lambda_j^{\mathcal{I}} t_i} sin(\lambda_j^{\mathcal{R}} t_i + \varphi_j) \qquad (1)$$

for $i = 1, \ldots, N$, where: $P_i = P(A_j, \lambda_j, \varphi_j; t_i)$ represents the disturbance pressure; A_j is the amplitude of the jth mode; $\lambda_j^{\mathcal{R}} = \omega_j/\Omega$ is the eigenfrequency of the jth mode; $\lambda_j^{\mathcal{I}}$ is the decay rate of the jth mode; φ_j is the phase of the jth mode and t_i is the time scaled by the rotation speed, Ω.

To estimate the parameters in a least squares sense, the sum of the squared differences between the calculated and observed responses is minimised, resulting in a set of nonlinear equations. Linearization is achieved by expanding each of the parameters in a Taylor series where only the linear terms are retained. The resulting set of simultaneous linear equations is solved iteratively until a criterion of convergence is satisfied. The variance of the parameter estimates is also available. This modelled sinusoid is then 'stripped' from the disturbance pressure record so that its residual can be analysed. This procedure is called *sequential fitting*. Implicit in the process of parameter estimation is the assumption that the residual is comprised of uncorrelated noise. Hence some bias will be inevitable if this assumption is violated.

The data was analysed by this linearized least squares (LLS) method because conventional time series analysis of quasi-stationary data can yield questionable results. It is our contention, based on our experience with our experimental data, that data segments having temporal extents comparable to the spin-down time for the fluid should be analysed. Modal interactions, which give rise to time-dependent decay rates (see SA Figure 9), require 'snapshots' to adequately portray the time evolution of the field. It is precisely for these reasons that the LLS method is used.

Results

The results that follow are divided into two groups. The first is concerned with retrograde modes while the second deals with prograde modes. Presented in what follows are representative results of this experimental study.

Retrograde Modes

Shown in Table 1 are the results for an inner spheroid tilt of $\epsilon = 8.0°$, an Ekman number$E \approx 9 \times 10^{-5}$, and forcing at $\lambda^{\mathcal{R}} = 1.000$. (The nondimensional Ekman number is $E = \nu/(\Omega R^2)$, and scales the viscous term in the Navier-Stokes equation [Greenspan, 1968].) The measured decay rate, which has been made dimensionless by multiplication with the spin-up time, appears in the first column. The measured dimensionless eigenfrequency, $\lambda^{\mathcal{R}} = \omega/\Omega$, is presented in the second column and the results are ordered in its decreasing value. Tentative modal identifications, based on frequency matching with the predicted

TABLE 1. Dimensionless Complex Eigenfrequencies for Forcing at $\lambda^{\mathcal{R}} = 1.000$, $\epsilon = 0.14$, $E \approx 8.8 \times 10^{-5}$.

Measured		Theoretical	mode
Decay Rate $\lambda^{\mathcal{I}}$	Frequency $\lambda^{\mathcal{R}}$	Frequency (sphere)	$n\ m\ k$
1.043±1.154	1.711±0.011	1.708	4 1 1
-1.626±2.490	1.572±0.023	1.588	8 2 1
0.925±0.407	1.229±0.004	1.306	6 2 1
0.550±0.006	1.127±0.006	1.046	5 2 1
1.756±0.064	1.008±0.001	1.000	2 1 1
0.543±0.528	0.812±0.005	0.790	7 3 1
0.215±0.602	0.637±0.005	0.612	4 2 1

values for the full sphere from the Poincaré equation, are given in the last two columns. (The modal numbering scheme (n, m, k) used here is that of AL. It is based on the period equation [Kudlick, 1966] where n is the degree of the Legendre Polynomial P_n^k, the smallest value of m corresponds to the simplest radial structure, and k is the azimuthal wavenumber.) The frequency shift due to the presence of the inner spheroid limits the identification for spatially complex modes. During free decay, as many as 7 modes were sequentially recoverable providing substantial evidence for interaction among modes. Negative decay rates indicating that the mode is growing in the data segment analysed, are thus not surprising. There is also strong evidence for indirect excitation as shown by the observation of modes having $\lambda^{\mathcal{R}} \geq 1$ in the table.

In Table 2, the results for the tilt $\epsilon = 11.1°$ of the inner spheroid are given. This increase in the magnitude of the perturbation significantly improves the amplitude of the response. Up to 12 modes can be recovered sequen-

TABLE 2. Dimensionless Complex Eigenfrequencies for Forcing at $\lambda^{\mathcal{R}} = 1.000$, $\epsilon = 0.19$, $E \approx 8.4 \times 10^{-5}$.

Measured		Theoretical	mode
Decay Rate $\lambda^{\mathcal{I}}$	Frequency $\lambda^{\mathcal{R}}$	Frequency (sphere)	$n\ m\ k$
-1.118±0.775	2.006±0.007	*	*
1.283±0.810	1.676±0.007	1.708	4 1 1
-2.785±0.897	1.355±0.008	1.353	10 3 1
4.070±0.928	1.267±0.008	1.306	6 2 1
-0.806±0.581	1.176±0.005	1.218	9 3 1
1.246±0.485	1.101±0.045	1.046	5 2 1
2.425±0.701	1.056±0.007	1.038	8 3 1
2.121±0.080	1.004±0.001	1.000	2 1 1
-1.816±0.435	0.883±0.004	0.856	10 4 1
4.061±3.228	0.799±0.029	0.790	7 3 1
9.202±1.049	0.772±0.009	0.728	12 5 1
-4.065±0.966	0.752±0.009	*	*

*Not identified as a low-order mode.

tially. Some of these are growing disturbances as shown by negative decay rates for the finite segment analysed. Further, the observations suggest that modes anywhere in the spectrum can be indirectly excited during free decay. The presence of the inner spheroid, and/or the degree of spatial complexity considered, has rendered some modes unidentifiable.

Studies of the forced response in the retrograde portion of the spectrum, revealed a second peak centered at $\lambda^{\mathcal{R}} \approx 0.740$. The results of free decay studies at this location are summarized in Table 3. Modal interaction, as witnessed by the recovery of up to 12 modes, is again present at this different setting of $\lambda^{\mathcal{R}}$. Indirectly, a suite of modes covering the full range of $\lambda^{\mathcal{R}}$ is excited during free decay.

TABLE 3. Dimensionless Complex Eigenfrequencies for Forcing at $\lambda^{\mathcal{R}} = 0.740$, $\epsilon = 0.19$, $E \approx 6.2 \times 10^{-5}$.

Measured		Theoretical	mode
Decay Rate $\lambda^{\mathcal{I}}$	Frequency $\lambda^{\mathcal{R}}$	Frequency (sphere)	$n\ m\ k$
2.110±2.759	1.396±0.022	1.353	10 3 1
2.042±1.944	1.261±0.017	1.306	6 2 1
0.499±0.693	1.112±0.005	1.046	5 2 1
2.439±1.906	1.016±0.015	1.000	2 1 1
0.273±0.442	0.986±0.003	*	*
0.314±0.785	0.910±0.006	*	*
-0.251±0.795	0.863±0.006	0.856	10 4 1
5.985±0.522	0.783±0.004	0.790	7 3 1
-0.264±0.196	0.760±0.002	0.728	12 5 1
1.598±0.363	0.686±0.003	0.633	9 4 1
-0.037±0.597	0.622±0.005	0.612	4 2 1
0.131±0.486	0.541±0.004	*	*

*Not identified as a low-order mode.

Prograde Modes

The results of Table 4 provide the first experimental verification of the existence of a suite of prograde modes. The largest resonance peak was centered at $\lambda^{\mathcal{R}} = -0.790$ and free decay studies were carried out there. We initially reported that prograde modes did *not* interact during free decay since only 1 mode with frequency near the excitation had been recovered. However, subsequent data analysis has shown that additional modes do exist so there is interaction as in the retrograde case.

TABLE 4. Dimensionless Complex Eigenfrequencies for Forcing at $\lambda^{\mathcal{R}} = -0.790$, $\epsilon = 0.19$, $E \approx 6.6 \times 10^{-5}$.

Measured		Theoretical	mode
Decay Rate $\lambda^{\mathcal{I}}$	Frequency $\lambda^{\mathcal{R}}$	Frequency (sphere)	$n\ m\ k$
-0.579±1.460	-1.225±0.012	-1.263	9 7 1
1.447±0.826	-1.054±0.007	-1.183	5 4 1
-1.093±0.321	0.992±0.003	1.000	2 1 1
1.505±0.780	-0.923±0.005	-0.893	10 7 1
6.487±0.540	-0.809±0.004	-0.820	4 3 1
-1.578±0.532	-0.752±0.004	-0.678	9 6 1

An interesting finding is the indirect excitation of the (retrograde) 'spin-over' or $(2,1,1)$ mode. An example of the recovery bias is evident in this value for the $(2,1,1)$ mode. Since this mode is very easy to excite [Greenspan, 1968], as it is the mode most strongly coupled to the forcing spheroid, the result is not surprising. In general, to clearly distinguish the indirect excitation of retrograde inertial waves by a prograde perturbation, or *vice-versa*, requires an array of pressure transducers which could be correlated in time. Equivalently, more than one superconducting gravimeter is necessary to distinguish prograde from retrograde waves in the Earth's fluid core.

Discussion

The principal results of this experimental investigation into the properties of inertial waves include verification of SA's claim for the existence of inertial waves in the spheroidal shell geometry. Comparison of SA's results for $\epsilon = 5.2°$ with those of the current series, $\epsilon = 8.0°$ and $\epsilon = 11.1°$, indicates that the interaction between modes during free decay increases with increasing tilt. Although Greenspan [1969] found nonlinear interaction to be insignificant with regard to coupling between modes, the present findings indicate that the interaction phenomenon appears to depend on nonlinear effects. (For a relative velocity scale, U, scaling the Navier-Stokes equation with $\epsilon = U/\Omega R$ (the Rossby number) shows that ϵ is a measure of the nonlinear effects.)

The steps taken to improve the resolution of the data, over that of the previous work [SA], has resulted in the recovery of significantly more modes. Free decay is observed to excite not just neighbouring modes (as claimed by SA) but a suite of modes. Indirectly, the entire spectrum is accessible during free decay. In addition, the first experimental evidence for a prograde spectrum, which has similar properties to its retrograde counterpart, has been presented.

The spherical geometry, at least at inertial periods, seems to be highly stable. Counterparts of resonance collapse, found in the cylindrical geometry [McEwan, 1970; Stergiopoulos and Aldridge, 1982; Thompson, 1970], have not been observed in the spherical shell. Malkus [1968] found turbulent flows in precessing an oblate spheroidal cavity. Though his experiments and ours are similar in nature, his findings relate to very small eigenvalues i.e., $\lambda^{\mathcal{R}} \rightarrow 0$.

The nature of the sequential recovery method is such that the parameters are estimated with the residual assumed to be uncorrelated noise. However, the residual can still contain decaying modes which acts like periodic noise. This, of course, invalidates the assumption and biases the estimates so obtained. This effect can be measured, for example, in the value obtained for the $(2,1,1)$ mode, since it is expected that the inner spheroid should *not* effect this mode.

The necessity of modelling the free decay in the interpretation of our experimental data has important applications to the processing of the LPG data. The nature of the LPG data is such that it is non-stationary or, at best, quasi-stationary. Conventional time series analysis can therefore produce misleading results. We are currently processing the LPG data, using the expertise gained from analysing the analogous experimental data, with these properties in mind.

The link between laboratory inertial wave investigations and the dynamics of the Earth's fluid outer core has been clearly demonstrated by AL. The laboratory counterparts of the (azimuthal number unity) modes that they identified in the LPG data of Melchior and Ducarme [1986] have been observed experimentally. The most significant property of these inertial waves is their interaction during free decay which gives rise to a suite of modes. This property, as suggested by AL, could be responsible for the observed rapid decay of those modes found in the LPG data. However, SA caution direct extrapolation of the laboratory results, having $E \sim \mathcal{O}(10^{-5})$, to the Earth's fluid core, $E \sim \mathcal{O}(10^{-15})$ [Gans, 1970]. They claim that the core's exceedingly low Ekman number brings the very existence of the inertial waves into question whereas, this 'degeneracy' might be removed at large Ekman number as in the case of the laboratory experiments.

It is our expectation that experimental investigations, into the effects of interactions, dissipation, and the possibility of resonance collapse, coupled with theoretical work, will ultimately lead to defining a more realistic fluid core. This will eventually help elucidate the role of the fluid core in the geodynamo and rotational geodynamics.

Acknowledgements. This research was supported by a generous operating grant from NSERC.

References

Aldridge, K. D., Axisymmetric inertial oscillations of a fluid in a rotating spherical shell, *Mathematika, 19,* 163-168, 1972.

Aldridge, K. D. and L. I. Lumb, Inertial waves identified in the Earth's fluid outer core, *Nature, 325,* 421-423, 1987.

Aldridge, K. D., L. I. Lumb and G. A. Henderson, Inertial modes identified in the Earth's fluid outer core, (this issue).

Aldridge, K. D. and A. Toomre, Axisymmetric inertial oscillations of a fluid in a rotating spherical container, *J. Fluid Mech., 37,* 307-323, 1969.

Dehant, V., Core undertones in an elliptical uniformly rotating Earth, (this issue).

Gans, R. F., Viscosity of the Earth's core, *J. geophys. Res., 77,* 360-366, 1970.

Greenspan, H. P., *The Theory of Rotating Fluids,* Cambridge University Press, 328 pp., 1968.

Greenspan, H. P., On the non-linear interaction between modes, *J. Fluid Mech., 36,* 257-264, 1969.

Gunn, J. S. and K. D. Aldridge, Inertial waves in a non-uniformly rotating fluid, (this issue).

Kudlick, M. D., On transient motions in a contained, rotating fluid, *Ph.D. thesis,* M.I.T., 1966.

Malkus, W. V. R., Precession of the Earth as the cause of geomagnetism, *Science, 160,* 259-264, 1968.

McEwan, A. D., Inertial oscillations in a rotating fluid cylinder, *J. Fluid Mech., 40,* 603-640, 1970.

Melchior, P. and B. Ducarme, Detection of inertial gravity oscillations in the Earth's core with a superconducting gravimeter at Brussels, *Phys. Earth planet. Int., 42,* 129-134, 1986.

Olson, P., A simple physical model for the terrestrial dynamo, *J. geophys. Res., 86,* 10875-10882, 1981.

Smylie, D. E., Variational calculation of wobble modes of the Earth, (this issue).

Smylie, D. E. and M. G. Rochester, Compressibility, core dynamics and the subseismic wave equation, *Phys. Earth planet. Int., 24,* 308-318, 1981.

Smylie, D. E. and M. G. Rochester, A variational principle for the subseismic wave equation, *Geophys. J. R. astr. Soc., 86,* 553-561, 1986.

Stergiopoulos, S. and K. D. Aldridge, Inertial waves in a fluid partially filling a cylindrical cavity during spin-up from rest, *Geophys. Astrophys. Fluid Dyn., 21,* 89-112, 1982.

Stergiopoulos, S. and K. D. Aldridge, Ringdown of inertial waves in a spheroidal shell of rotating fluid, *Phys. Earth planet. Int., 36,* 17-26, 1984.

Stewartson, K. and J. A. Rickard, Pathological oscillations of a rotating fluid, *J. Fluid Mech., 35,* 759-773, 1969.

Szeto, A. M. K. and D. E. Smylie, Coupled motions of the inner core and possible geomagnetic implications, *Phys. Earth planet. Int., 36,* 27-42, 1984.

Thompson, R., Diurnal tides and shear instabilities in a rotating cylinder, *J. Fluid Mech., 40,* 737-751, 1970.

THE EXCITATION OF CORE MODES BY EARTHQUAKES

David J. Crossley

Geophysics Laboratory, McGill University, 3450 University Street
Montreal, P.Q., Canada H3A 2A7

Abstract. Prediction of the theoretical spectrum of the internal motions of a fluid-filled rotating spherical shell, with application to the Earth's core, has recently become a pressing issue in physics of the Earth's deep interior. The recent claim that peaks in the observational record from the Brussels superconducting gravimeter are associated with inertial waves in the Earth's core raises many interesting questions concerning the calculation, identification, excitation and damping of core modes.

To date in core dynamics the concern has been with the free-mode eigenspectrum, which generally has to be solved as a preliminary to any forced problem. Considerable progress is being made on the calculations for certain models of density stratification and for certain period ranges of wave motions in the core, though results for an arbitrary core stratification at all periods are still incomplete. Nevertheless, the forced excitation problem can be solved approximately using standard normal mode excitation theory. The goal is to determine whether a large earthquake can excite core modes that generate perturbations in gravity potential at the Earth's surface with the approximate amplitudes of the spectral peaks observed at Brussels.

This process of summing a large number of spheroidal modes throughout the Earth is computationally prohibitive unless simplifying assumptions are made. It is reasonable to ignore Coriolis coupling (and also ellipticity, at least to this order of approximation) in the shell and inner core due to the high bulk rigidity of these regions. In the fluid core it is important to retain Coriolis coupling, though necessary to truncate the spherical- harmonic expansion after a number of terms. Results are presented for a sample earthquake source in a non-rotating Earth model and indicate that the excitation of a single mode is well below the current observational threshold. However in future calculations with an extended number of modes some enhancement of the response may be expected.

Introduction

The calculation of the theoretical spectrum of hydrodynamic waves in the Earth's core has enjoyed, since Pekeris and Accad [1972], a slow but steady development. The topic has until recently been considered (by all but a few devotees) as a low-priority project, simply on the general view that the Earth's core must be convecting vigorously enough (by gravitational separation) to maintain the geodynamo and hence upset the conditions for internal gravity waves to exist. It is widely believed that the existence of core modes depends on the fluid being stably stratified. This is not strictly true if one includes inner core translation, inertial waves and Rossby waves as core modes, though of course the dynamics of a neutrally stratified core are not as interesting as one with regions of stablity.

To the question "is the outer fluid core stably stratified ?" a seismologist would answer "no, the density stratification is indistinguishable from the Adams-Williamson condition (neutral stability) according to body wave and normal mode data; any departures are simply noise". The dynamo theorist would answer "no, it can't be, otherwise the dynamo would not operate"; the thermodynamicist would reply "no, there's not enough energy to maintain a sub-adiabatic gradient". On the other hand the applied mathematician takes the viewpoint "that's not for me to decide; in any case why not consider the equations for a thick shell of rotating fluid as a model for a core with an arbitrary stability profile ?". I refer the reader to Masters [1979], Verhoogen [1981; Ch.4], Stacey [1981] and Friedlander [1985] for examples of these viewpoints.

Until recently theoretical studies (e.g. Crossley [1984], Smylie and Rochester [1986]) have been largely unconstrained by observations, the seismological data being sufficiently inconclusive on the question of core adiabaticity. But since the publication of the superconducting gravimeter record from Brussels [Melchior and Ducarme, 1986;

Zürn et al., 1987], we must add, to the above, the geodesist's answer "maybe; since we now have appropriate data of unprecedented sensitivity, this should be examined for the direct effects of core motions". For indeed the challenge has now been laid down to establish, not yet the extent to which the core departs from adiabaticity (perhaps the ultimate goal), but the more immediate question of whether for any conceivable core model it it possible to generate observable gravity signals at the Earth's surface from internal harmonic motions of the fluid.

Despite the complexity of the theoretical question, it is important to respond to the geodetic challenge as soon as possible, which is the purpose of this paper. We therefore shall leave aside a review of the core mode studies *per se*, referring to Crossley [1984] and Smylie and Rochester [1986] for representative status reports, and concentrate on the excitation problem directly.

Excitation in a Non-rotating Earth Model

The forced solution

We represent the excited normal mode problem by the equation

$$(L + \partial_t^2)\mathbf{S}(\mathbf{r}, t) = \mathbf{f}(\mathbf{r}, t) \tag{1}$$

where L is the elastic-gravitational operator for normal modes of a realistic Earth model, $\mathbf{S}$ is the total Lagrangian displacement in that model and $\mathbf{f}$ is the forcing term, all functions of position $\mathbf{r}$ and time t. The main boundary conditions are that of a stress-free surface and a harmonic form for the gravitational potential there. Additionally, there are boundary conditions at the Earth's centre and at internal discontinuities which depend on the details of the Earth model. The forcing $\mathbf{f}$ can be either external (e.g. tidal, impact) or internal (e.g. earthquake). Development of the theory follows that given by Gilbert [1970] as extended by Dahlen and Smith [1975] and vindicated by Chao [1982].

The solution to (1) is

$$\mathbf{S}(\mathbf{r}, t) = \sum_n \mathbf{u}_n(\mathbf{r}) \int_0^t \frac{1}{\omega_n} \sin \omega_n(t - \tau) f_n(\tau) d\tau \tag{2}$$

for $t \geq 0$, where

$$f_n(t) = (\mathbf{u}_n^*, \mathbf{f}) = \int_V \mathbf{u}_n^*(\mathbf{r}) \cdot \mathbf{f}(\mathbf{r}, t)\rho_0 dV \tag{3}$$

and the integration is over a volume V of the model with density ρ_0. The displacements $\mathbf{u}_n$ are (complex) free-mode eigenfunctions of the homogenous system equivalent to (1), each associated with an eigenfrequency ω_n:

$$(L - \omega_n^2)\mathbf{u}_n(\mathbf{r})e^{i\omega_n t} = 0. \tag{4}$$

It is understood that the pair of variables $(\omega_n, \mathbf{u}_n)$ is called a "mode"; for example in a non-rotating, spherical, Earth model each mode can be represented by a single spherical harmonic function. For an earthquake excitation we take the forcing to be a (Heaviside) unit step in time

$$\mathbf{f}(\mathbf{r}, t) = \mathbf{f}(\mathbf{r})H(t). \tag{5}$$

The solution (2) can then be expressed as a sum over all forced eigenfunctions,

$$\mathbf{S}(\mathbf{r}, t) = \sum_n \mathbf{s}_n(\mathbf{r})(1 - \cos \omega_n t) \tag{6}$$

each with a spatial part which is simply a scaled form of the free-mode eigenfunction

$$\mathbf{s}_n(\mathbf{r}) = \frac{1}{\omega_n^2}(\mathbf{u}_n^*, \mathbf{f}_\infty)\mathbf{u}_n(\mathbf{r}). \tag{7}$$

In (7), $\mathbf{f}_\infty$ signifies that the force distribution is assumed to be maintained indefinitely into the future. As Gilbert [1970] shows, for times $t \gg T$ (the duration of the earthquake), the response (6) can be modified by including an assumed damping of each mode

$$\omega_n \longrightarrow \omega_n(1 + \frac{i}{2Q_n}), \qquad Q_n \gg 1, \tag{8}$$

according to its Q value. We note from (7) that the amplitude of an individual mode is proportional to $1/\omega_n^2$, which therefore weights more heavily the long-period response of the Earth, i.e. the gravity (core) modes, compared to the elastic (seismic) modes. As will become clear later, other factors of course also contribute to the relative excitation of the two families of modes.

In order to evaluate (7) it is necessary to establish an absolute level for the free-mode eigenfunctions $\mathbf{u}_n$ and this is done through the requirement that any two such eigenfunctions be orthonormal over the parameter space of the model :

$$(\mathbf{u}_i^*, \mathbf{u}_j) = \int_V \mathbf{u}_i^*(\mathbf{r}) \cdot \mathbf{u}_j(\mathbf{r})\rho_0 dV = \delta_{ij}. \tag{9}$$

Assume for example that we take $\mathbf{u}_n(\mathbf{r})$ to be a (finite) sum of spheroidal oscillations of different azimuthal numbers (i.e. a multiplet), all of which have the same eigenfrequency in a non-rotating Earth model, i.e.

$$\mathbf{u}_n(\mathbf{r}) = \sum_{m=-l}^{l} \mathbf{S}_l^m(r, \theta, \phi) \,. \tag{10}$$

Each spheroidal vector function

TABLE 1. Boundary Conditions for Spheroidal and Torsional Terms.
$a_{1\dots6}$ are Free Constants and [ ] are Boundary Conditions to be Applied.

	centre $r=0$	icb $r=a$	cmb $r=b$	surface $r=d$
(a) spheroidal				
radial displacement	$y_1=0$			
radial stress	$y_2=a_1$			$[y_2=0]$
transverse displacement	$y_3=0$		$y_3=a_4$	
transverse stress	$y_4=a_2$	$[y_4=0]$	$y_4=0$	$[y_4=0]$
gravitational perturbation	$y_5=0$			
gravitational flux	$y_6=a_3$			$[y_6=-(l+1)y_5/d]$
(b) torsional				
transverse displacement	$y_7=0$		$y_7=a_6$	
transverse stress	$y_8=a_5$	$[y_8=0]$	$y_8=0$	$[y_8=0]$

$$\mathbf{S}_l^m(r,\theta,\phi)=(u_l^m P_l^m \hat{\mathbf{r}}+v_l^m\frac{dP_l^m}{d\theta}\hat{\theta}+\frac{im}{r\sin\theta}v_l^m P_l^m\hat{\phi})e^{im\phi} \tag{11}$$

is expressed in terms of an Associated Legendre Function $P_l^m(\cos\theta)$ and two radial coefficients $u_l^m(r)$ and $v_l^m(r)$. In this case, the orthonormality condition (9) applies to each spherical harmonic independently. Because a gravimeter at the Earth's surface responds only to spheroidal displacements, we can ignore the torsional displacements in the non-rotating case.

Boundary Conditions

For a spheroidal eigenfunction of degree l, the solution requires a 6'th order set of ODE's in the inner core and shell and a 4'th order solution in the fluid outer core. Integrating through the model we have the usual set of boundary conditions and free constants a_i, using the notation of Alterman et al. [1959], shown in Table 1. For the free modes, at an eigenfrequency the determinant of all trial solutions $y_i^{(j)}$ must vanish to satisfy the boundary conditions. The undetermined constants a_i can be determined uniquely for each of the trial solutions by solving the orthonormality requirement (9) together with the surface ($r=d$) conditions $[i,j=1\dots4]$

$$a_i a_j E_{ij} = \delta_{ij} \tag{12}$$

$$\begin{aligned} a_j y_2^{(j)} &= 0 \\ a_j y_4^{(j)} &= 0 \\ a_j(y_6^{(j)}+\frac{(l+1)}{d}y_5^{(j)}) &= 0. \end{aligned} \tag{13}$$

In (12) $\omega^2 E_{ij}$ is just the kinetic energy of a single mode

$$E_{ij}=\frac{4\pi}{2l+1}\frac{(l+m)!}{(l-m)!}\int_0^d[y_1^{(i)}y_1^{(j)}+l(l+1)y_3^{(i)}y_3^{(j)}]\rho r^2dr. \tag{14}$$

Equations (12) and (13) are nonlinear in the unknown constants a_j and have to be solved iteratively. The final $\mathbf{u}_n$ can then be found by reconstruction of the radial function from the trial solutions

$$y_i(r)=a_j y_i^{(j)} \tag{15}$$

Excitation Coefficient

According to theory developed for the excitation of the Chandler wobble by an earthquake source [Smylie and Mansinha, 1971; Dahlen, 1971], the force term can be reasonably approximated as a point vector formally equivalent to a unit moment tensor. For an earthquake of magnitude M, we expand the force as sum over all degrees l and orders m of spheroidal and toroidal vectors

$$\mathbf{f}_\infty=M\sum_{l=0}^{\infty}\sum_{k=-l}^{l}(\mathbf{S}_l^k+\mathbf{T}_l^k). \tag{16}$$

Suppose these vectors have radial functions $u_l^{\prime k}, v_l^{\prime k}$ for spheroidal and $t_l^{\prime k}$ for torsional components, associated with a spherical harmonic function in the usual way. For convenience, these coefficients are also listed in Table 2.

We find the scalar excitation coefficient for a spheroidal mode to be

$$(\mathbf{u}_n^*,\mathbf{f}_\infty)=M\sum_m\frac{4\pi}{2l+1}\frac{(l+m)!}{(l-m)!}\cdot \int_0^d[u_l^m u_l^{\prime m}+l(l+1)v_l^m v_l^{\prime m}]\rho_0 r^2dr, \tag{17}$$

TABLE 2. Radial Coefficients of Unit Point Force Excitation. All Terms Have to be Multiplied by the Common Factor $(2l+1)/8\pi r^3$. Source Radius r_0, Colatitude 0°, Dip Angle α, $\delta = \delta(r - r_0)$, $\delta' = \partial_r \delta$.

m	Strike Slip	Dip Slip
0	$u_l' = 0$	$u_l' = 2(r\delta' + \delta)\ \sin 2\alpha$
	$v_l' = 0$	$v_l' = -\delta\ \sin 2\alpha$
	$t_l' = 0$	$t_l' = 0$
1	$u_l' = -\delta\ \cos\alpha$	$u_l' = -i\delta\ \cos 2\alpha$
	$v_l' = [(r\delta' + \delta)/l(l+1)]\ \cos\alpha$	$v_l' = i[(r\delta' + \delta)/l(l+1)]\ \cos 2\alpha$
	$t_l' = -i[(r\delta' + \delta)/l(l+1)]\ \cos\alpha$	$t_l' = [(r\delta' + \delta)/l(l+1)]\ \cos 2\alpha$
-1	$u_l' = l(l+1)\delta\ \cos\alpha$	$u_l' = -il(l+1)\delta\ \cos 2\alpha$
	$v_l' = -(r\delta' + \delta)\ \cos\alpha$	$v_l' = i(r\delta' + \delta)\ \cos 2\alpha$
	$t_l' = -i(r\delta' + \delta)\ \cos\alpha$	$t_l' = -(r\delta' + \delta)\ \cos 2\alpha$
2	$u_l' = 0$	$u_l' = 0$
	$v_l' = i[\delta/l(l+1)]\ \sin\alpha$	$v_l' = -[\delta/2l(l+1)]\ \sin 2\alpha$
	$t_l' = [\delta/l(l+1)]\ \sin\alpha$	$t_l' = i[\delta/2l(l+1)]\ \sin 2\alpha$
-2	$u_l' = 0$	$u_l' = 0$
	$v_l' = -i(l-1)(l+2)\delta\ \sin\alpha$	$v_l' = -[(l-1)(l+2)\delta/2]\ \sin 2\alpha$
	$t_l' = (l-1)(l+2)\delta\ \sin\alpha$	$t_l' = -i[(l-1)(l+2)\delta/2]\ \sin 2\alpha$

where the radial functions $u_l'^m, v_l'^m$ are as derived in Smylie and Mansinha [1971] and Mansinha et al. [1979]. Note that a mode here is uniquely identified with a single spheroidal oscillation of degree l.

As an example, let us take a dip-slip earthquake source of moment M for a mode with azimuthal number $m = 2$. From Table 2

$$u_l'^2 = 0$$

$$v_l'^2 = \frac{2l+1}{8\pi r^2 l(l+1)}[-\frac{1}{2r}\delta(r - r_0)]\ \sin 2\alpha, \qquad (18)$$

whence the excitation coefficient is

$$(\mathbf{u}_n^*, \mathbf{f}_\infty) = -\frac{(l+m)!}{(l-m)!}\frac{\rho_0 y_3(r_0)}{4r_0} M \sin 2\alpha, \qquad (19)$$

where r_0 is the Earth radius at which the point force is applied, α is the dip and y_3 is the transverse displacement radial function at that radius. Then using (7) the forced eigenfunction has the form

$$\mathbf{s}_n(\mathbf{r}) = -\frac{(l+m)!}{(l-m)!}\frac{\rho_0 y_3(r_0)}{4r_0\omega_n^2} M \sin 2\alpha\ \mathbf{u}_n(\mathbf{r}). \qquad (20)$$

It is worth commenting at this stage that the effect of source depth $(1/r_0)$ is much less significant than that of seismic moment M, especially as the free-mode radial functions (e.g. y_3 in the above expression) are almost constant with radius in the crust of the Earth. On the other hand the seismic moment scales the eigenfunctions $\mathbf{u}_n$ in (20) directly.

Surface Gravity

As indicated by Johnson and Smylie [1977], the response of a gravimeter at the Earth's surface can be calculated by summing three contributions :

(i) displacement through the static field.

$$g_1 = y_1(4\pi G\rho_0 - \frac{2}{r}g_0)$$

(ii) inertial acceleration of the ground.

$$g_2 = \omega^2 y_1$$

(ii) perturbation of the gravitational attraction.

$$g_3 = -\frac{dy_5}{dr}$$

where y_1 and y_5 are the radial functions of the forced eigenfunctions at the surface, i.e. the radial functions of $\mathbf{s}_n(\mathbf{r})$ in equation (20). We need of course to sum these for the total gravity perturbation g_t (Table 3).

TABLE 3. Sample Dip-Slip Earthquake Excitation.
Source - Dip Angle 10°; $M = 10^{30}$ dyne cm; Depth 25 km.

Model	n	period	gravity($\mu gals$) g_1	g_2	g_3	g_t	displacement (cm) $y_1(d)$	$y_1(0)$
(a) spheroidal modes ${}_nS_2^2$								
B1-6	5	478s	-0.70 10^{-4}	0.16 10^{-1}	-0.21 10^{-3}	0.16 10^{-1}	0.95 10^{-4}	0.
	4	581s	0.92 10^{-4}	-0.15 10^{-1}	0.27 10^{-3}	-0.14 10^{-1}	-0.12 10^{-3}	0.
	3	905s	-0.17 10^{-3}	0.11 10^{-1}	-0.58 10^{-3}	0.11 10^{-1}	0.23 10^{-3}	0.
	2	1081s	-0.20 10^{-4}	0.90 10^{-3}	-0.61 10^{-4}	0.82 10^{-3}	0.27 10^{-4}	0.
	1	1469s	0.26 10^{-3}	-0.64 10^{-2}	-0.47 10^{-3}	-0.66 10^{-2}	-0.35 10^{-3}	0.
	0	3232s	0.95 10^{-4}	-0.48 10^{-3}	0.15 10^{-4}	-0.37 10^{-3}	-0.13 10^{-3}	0.
	-1	9.56h	-0.14 10^{-6}	0.65 10^{-8}	-0.40 10^{-6}	-0.53 10^{-6}	0.19 10^{-6}	0.
	-2	15.66h	-0.75 10^{-7}	0.13 10^{-8}	-0.17 10^{-6}	-0.24 10^{-6}	0.10 10^{-6}	0.
	-3	22.66h	-0.33 10^{-7}	0.26 10^{-9}	-0.93 10^{-7}	-0.13 10^{-6}	0.44 10^{-7}	0.
	-4	30.53h	-0.23 10^{-7}	0.10 10^{-9}	-0.55 10^{-7}	-0.78 10^{-7}	0.32 10^{-7}	0.
	-5	38.17h	-0.20 10^{-7}	0.57 10^{-10}	-0.58 10^{-7}	-0.78 10^{-7}	0.27 10^{-7}	0.
	-6	44.52h	-0.92 10^{-8}	0.19 10^{-10}	-0.20 10^{-7}	-0.29 10^{-7}	0.12 10^{-7}	0.
B1-12	-1	19.01h	-0.36 10^{-7}	0.41 10^{-9}	-0.97 10^{-7}	-0.13 10^{-6}	0.48 10^{-7}	0.
	-2	31.65h	-0.19 10^{-7}	0.79 10^{-10}	-0.42 10^{-7}	-0.61 10^{-7}	0.26 10^{-7}	0.
	-3	44.67h	-0.10 10^{-7}	0.21 10^{-10}	-0.28 10^{-7}	-0.38 10^{-7}	0.13 10^{-7}	0.
B1-18	-1	28.28h	-0.17 10^{-7}	0.85 10^{-10}	-0.44 10^{-7}	-0.61 10^{-7}	0.22 10^{-7}	0.
	-2	47.19h	-0.93 10^{-8}	0.17 10^{-10}	-0.20 10^{-7}	-0.30 10^{-7}	0.13 10^{-7}	0.
(b) translational modes ${}_nS_1^1$								
B1-6	1	2473s	0.24 10^{-2}	-0.21 10^{-1}	0.73 10^{-2}	-0.11 10^{-1}	-0.32 10^{-2}	-0.37 10^{-1}
	-1	12.18h	0.29 10^{-3}	-0.79 10^{-5}	0.65 10^{-3}	0.93 10^{-3}	-0.39 10^{-3}	0.24
	-2	23.96h	0.38 10^{-3}	-0.27 10^{-5}	0.24 10^{-3}	0.62 10^{-3}	-0.52 10^{-3}	-0.32
	-3	36.94h	0.42 10^{-2}	-0.13 10^{-4}	-0.30 10^{-2}	0.12 10^{-2}	-0.56 10^{-2}	0.58

Sample Solutions

We show several representative normal modes of a non-rotating Earth model with a stably stratified core ($2\pi/N$ = 6 hr, N = buoyancy frequency). In Figures 1(a)-(c), we show radial functions for several $l = 2, m = 2$ seismic modes with radial numbers +4, +2 and 0 and core modes with radial numbers -1 and -3 in Figures 1(d)-(e). The eigenfunction normalisations are also shown; a modified B1 model of Jordan and Anderson [1974] was used. Naturally, the eigenfunctions in the shell for the core modes in Figures 1(d)-(e) are much reduced over those of the seismic overtones in Figures 1(a)-(c). We also show, in Figures 2(a)-(c), three inner core translational modes, the first for overtones $n = +1$, and the next two for undertones $n = -1, -2$. For these modes the inner core displacement is almost constant, as expected for a pure translation.

The forced eigenfunctions were calculated for a suite of overtone and undertone modes, including the modes shown in the figures, and the results are shown in Table 3 for parameters similar to the Chilean event of 1960 [Smith, 1976]. The designation B1-6, B1-12 and B1-18 denotes modified B1 models with core buoyancy periods ranging from 6 hr (strongly stable core) through 12 hr (critically stable core) to 18hr (weakly stable core). It can be seen that the eigenfunctions for $n = -1, -2$ are almost identical for all three Earth models. In Table 3 the three components of the gravity signal have been given separately; the final two columns give the amplitude of radial displacement at the Earth's surface and centre respectively.

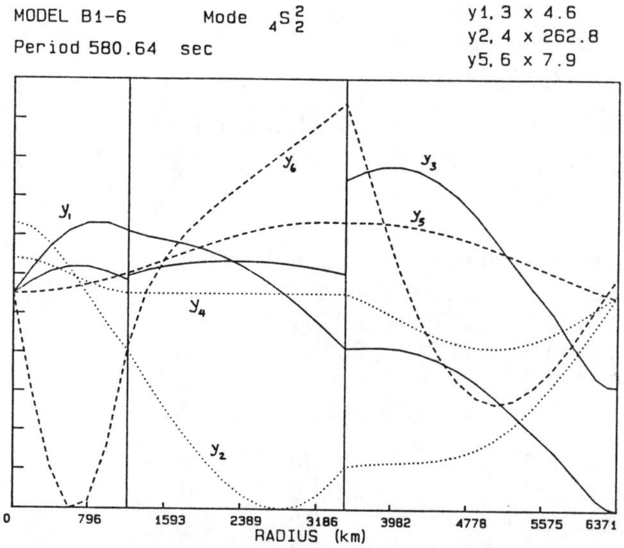

Fig. 1(a). Radial functions for normal mode overtone ${}_4S_2^2$ for degree $l = 2$, order $m = 2$ for Earth model B16 with a strongly stable core. The eigenfunctions are normalised to the relative amplitudes given at the top right of the figure.

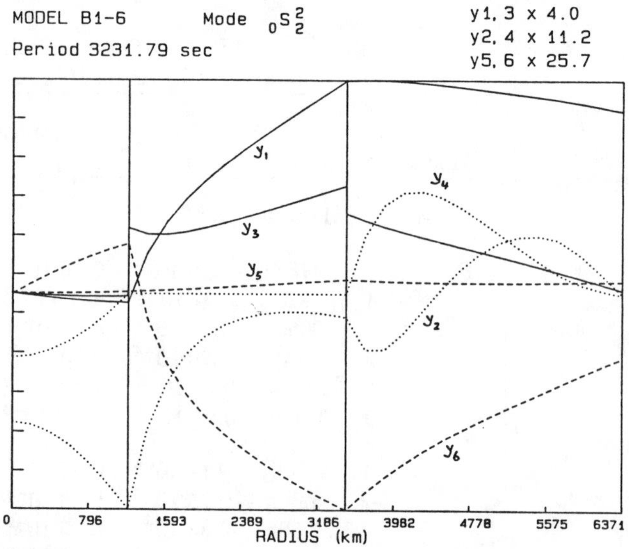

Fig. 1(c). As Fig. 1(a) for the fundamental normal mode ${}_0S_2^2$.

It has been conjectured [Crossley, 1984] that the large rigidity contrast across the core mantle boundary might imply that core modes could be excited with only a small shell displacement. As argued also by Smylie [personal communication], this would be the case if the core and shell traded off kinetic and potential energies respectively for the core modes. We must remember that the equipartition of kinetic and potential (elastic + gravitational) energy for a mode is only required for the whole Earth, not for each region individually. A good example of this is demonstrated by mode ${}_0S_2^2$ in Table 4 where the elastic potential energy of the shell is traded off against gravitational energy in the outer core. To test the above, the kinetic, elastic and gravitational potential energies were computed for some modes of model B16 using the expressions given by Kovach and Anderson [1967].

The results, in Table 4, show the above speculation to

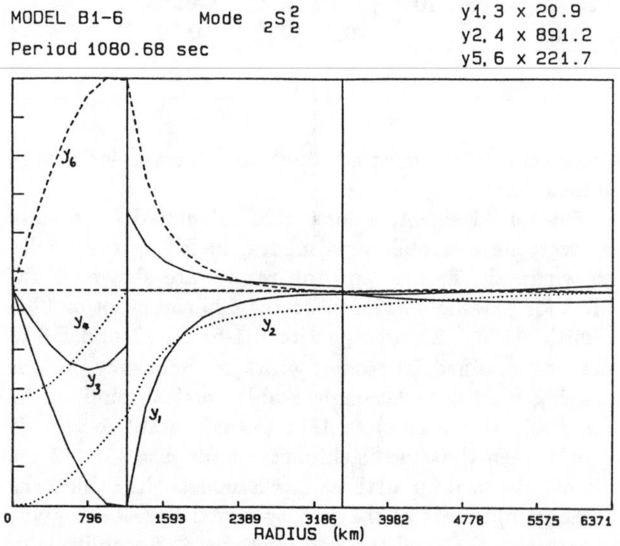

Fig. 1(b). As Fig. 1(a) for normal mode overtone ${}_2S_2^2$ showing primarily inner core motion.

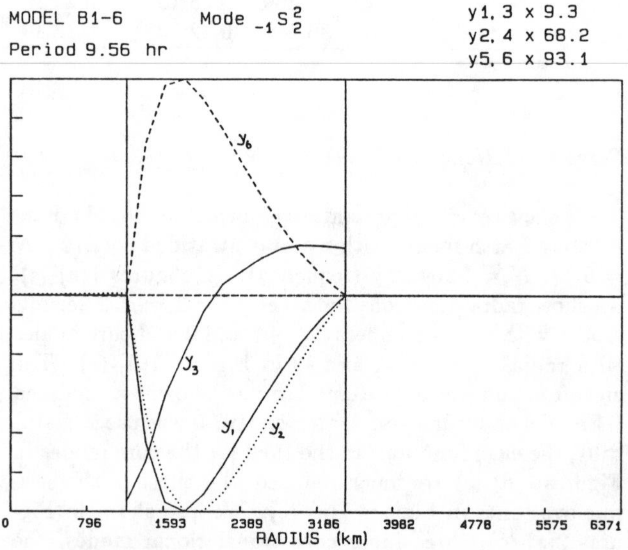

Fig. 1(d). As Fig. 1(a) but for normal mode undertone ${}_{-1}S_2^2$ showing dominance of motion in the core.

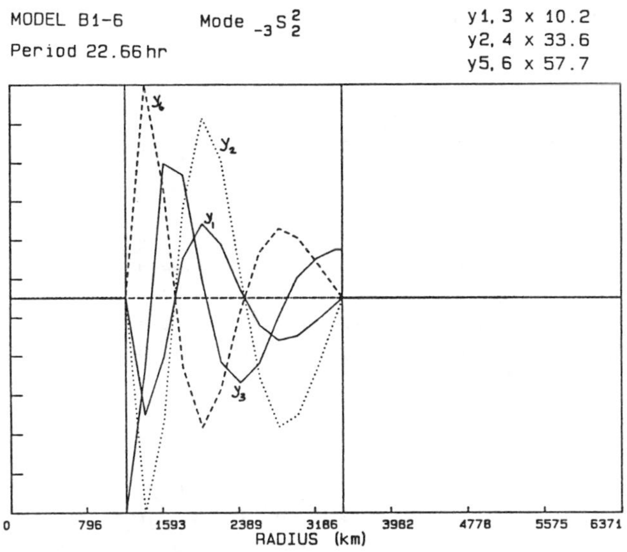

Fig. 1(e). As Fig. 1(d) but for normal mode undertone $_{-3}S_2^2$.

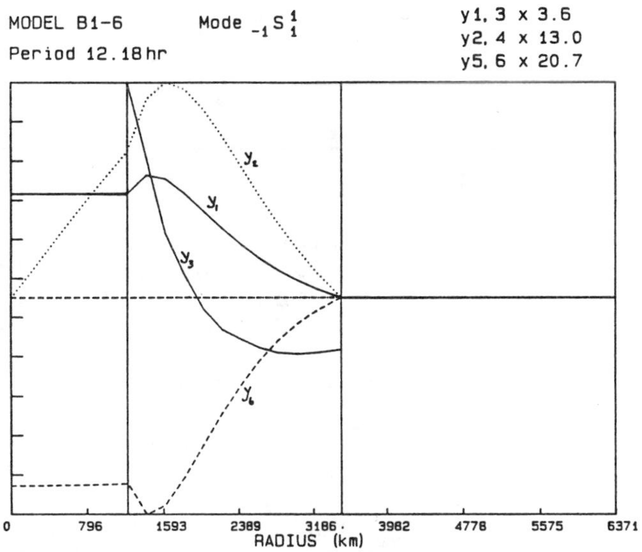

Fig. 2(b). As Fig. 2(a) but for Slichter mode $_{-1}S_1^1$ showing translation for the inner core only.

be unfounded. The table entries are the fraction of kinetic, elastic and gravitational potential energies in the inner core, outer core and shell for each mode (summing across all the KE, PE and GE columns for each row gives unity within numerical error). Equipartition for each mode is ensured by KE=PE+GE.

It can be seen that the regular spheroidal seismic modes spread energy throughout the Earth. Apart from unusual modes such as $_2S_2^2$ (which is predominantly an inner core mode), the shell contains the bulk of the energy. For the core modes, both kinetic and potential energies are almost exclusively confined to the outer core. One interesting feature is the increase in potential energies in the outer core as N^2 is decreased. For the long period translational modes of the inner core, the shell contains almost no energy and the inner and outer cores trade off large contributions of gravitational energy. Thus core modes and translational modes of the inner core are unlikely to

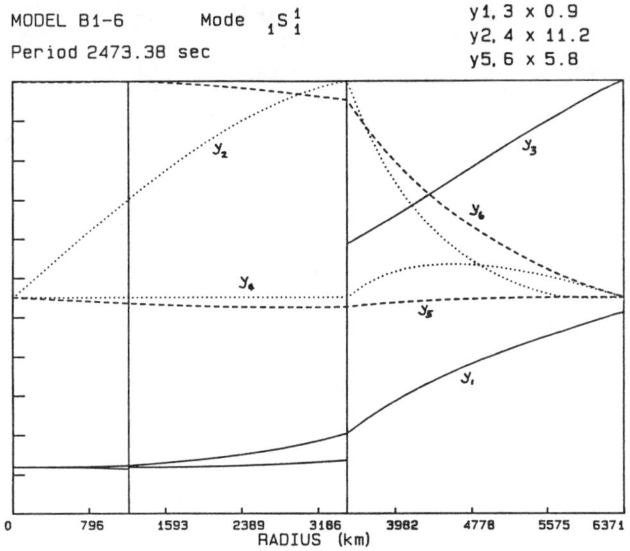

Fig. 2(a). Radial functions for Slichter normal mode overtone $_1S_1^1$ for degree $l = 1$, order $m = 1$ for Earth model B16 with a strongly stable core. The inner and outer core displacements are virtually constant indicating rigid body translation.

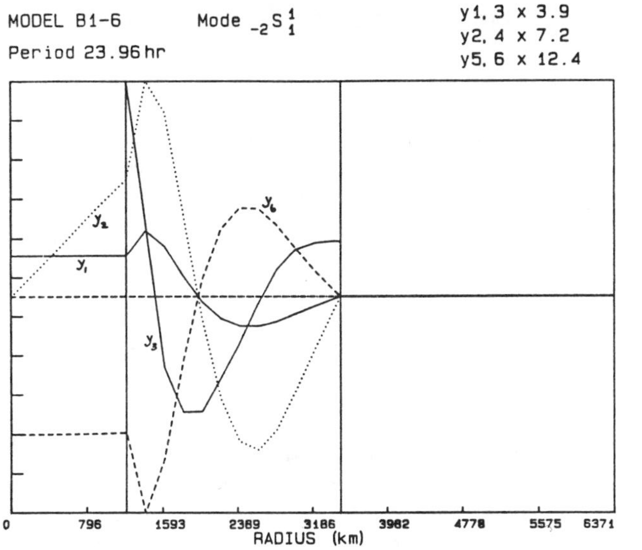

Fig. 2(c). As Fig. 2(b) but for Slichter mode $_{-2}S_1^1$.

TABLE 4. Energy in Spheroidal Modes[1]. KE = Kinetic Energy, PE = Elastic Potential Energy, GE = Gravitational Potential Energy.

		inner core			outer core			shell		
Model	n	KE	PE	GE	KE	PE	GE	KE	PE	GE
(a) spheroidal modes ${}_nS_2^2$										
B1-6	5	0.0096	0.0050	-0.0	0.1296	0.1955	-0.0083	0.8608	0.8110	-0.0056
	4	0.0027	0.0015	-0.0	0.0454	0.0790	-0.0045	0.9519	0.9571	-0.0334
	3	0.0552	0.0674	0.0056	0.3941	0.2352	0.0076	0.5507	0.7337	-0.0496
	2	0.5183	0.9150	0.1041	0.4294	0.0688	-0.1334	0.0523	0.0532	-0.0077
	1	0.0011	0.0036	0.0005	0.1243	0.0159	0.0376	0.8746	0.9650	-0.0224
	0	0.0	0.0006	0.0003	0.3878	0.0002	0.3697	0.6122	0.6562	-0.0269
	-1	0.0	0.0002	0.0002	1.0000	6.3533	-5.3534	0.0	0.0004	-0.0003
	-2	0.0	0.0001	0.0001	1.0000	6.9958	-5.9952	0.0	0.0002	-0.0002
	-3	0.0	0.0001	0.0	1.0000	6.9926	-5.9907	0.0	0.0001	-0.0001
	-4	0.0	0.0	0.0	1.0000	7.8956	-6.8916	0.0	0.0001	-0.0001
	-5	0.0	0.0	0.0	1.0000	8.5191	-7.5134	0.0	0.0001	-0.0
	-6	0.0	0.0	0.0	1.0000	6.3204	-5.3188	0.0	0.0	-0.0
B1-12	-1	0.0	0.0002	0.0002	1.0000	25.1509	-24.1454	0.0	0.0001	-0.0001
	-2	0.0	0.0	0.0	1.0000	28.1515	-27.1361	0.0	0.0001	-0.0
	-3	0.0	0.0	0.0	1.0000	27.6728	-26.6432	0.0	0.0	-0.0
B1-18	-1	0.0	0.0	0.0	1.0000	55.8194	-54.7914	0.0	0.0001	-0.0
	-2	0.0	0.0	0.0	1.0000	62.7792	-61.7012	0.0	0.0	-0.0
(b) translational modes ${}_nS_1^1$										
B1-6	1	0.0224	0.0002	0.0052	0.4016	0.0510	0.1216	0.5760	0.9309	-0.1089
	-1	0.1397	0.2181	23.3475	0.8603	6.0008	-28.5670	0.0	0.0006	-0.0008
	-2	0.0272	0.0169	17.7571	0.9728	7.4704	-24.4004	0.0	0.0002	-0.0004
	-3	0.0067	0.1004	10.4388	0.9933	7.3624	-16.9119	0.0	0.0001	-0.0004

[1] Normalised so total KE = $\omega^2 E_{ii} = 1$, see equation (14)

be damped by mantle elasticity, for example; conversely, only a very large mantle source can excite them.

We may summarise the results as follows :

- The overall excitation of the core modes is down some 4-6 orders of magnitude compared to the seismic normal modes due to
  1. the extremely small mantle displacements
  2. the very small radial derivative of gravity perturbation near the surface

  These results are qualitatively similar to the calculation of core mode excitation by Alterman et. al [1959], though of course the periods here are quite different from the 100 minute oscillation of their model.

- There is only a second order effect of excitation with core stability and mode frequency. Although we cannot here calculate what the excitation would be for a neutral ($N = 0$) core, inertial waves, being purely toroidal, would not be expected to produce surface gravity effects directly [*cf.* Aldridge and Lumb, 1987].

- The inner core translations yield results very similar to Smith [1976]. These modes generally have a better chance of being detected than the core modes. In addition the longer period translations are just as energetic as the 'short' period motion with $n = +1$, indicating the preferential detection of these motions over the core modes. On the other hand, attempts to find evidence of these inner core motions

TABLE 5. Options for Numerical Solution. Each Spheroidal Term Requires 6 ODE's in a Solid and 4 in a Fluid. Each Torsional Term Requires 2 ODE's in a Solid.

inner core	outer core	shell
Best[1]		
$S_l^m + T_{l+1}^m + S_{l+2}^m + \ldots$	$S_l^m + T_{l+1}^m + S_{l+2}^m + \ldots$	$S_l^m + T_{l+1}^m + S_{l+2}^m + \ldots$
$N^4 = 80$	$N = 40$	$N = 80$
Okay[2]		
$S_l^m,\ S_{l+2}^m, \ldots$	$S_l^m + S_{l+2}^m + \ldots$	$S_l^m,\ S_{l+2}^m, \ldots$
$N = 60$	$N = 40$	$N = 60$
Poor[3]		
S_l^m	$S_l^m + S_{l+2}^m + \ldots$	S_l^m
$N = 6$	$N = 40$	$N = 6$

[1] Full coupling in solid and fluid region
[2] Torsional terms ignored in solid regions
[3] All but one spheroidal term ignored in solid regions
[4] Total number of ODE's assuming 10 spherical harmonics

with a spring gravimeter in Antartica have so far failed [Jackson and Slichter, 1974].

Extension to a Rotating Earth Model

Following Chao [1982], the modification to (1) for a rotating Earth is straightforward and leads to

$$(L + C\partial_t + \partial_t^2)\mathbf{S}(\mathbf{r}, t) = \mathbf{f}(\mathbf{r}, t) \qquad (21)$$

where $C = 2\Omega$ x is the Coriolis operator and L now includes the effect of the centrifugal potential. One difficulty that appears immediately in the new normalisation condition

$$(\omega_m + \omega_n)(\mathbf{u}_m^*, \mathbf{u}_n) - 2i(\mathbf{u}_m^*, \mathbf{\Omega} \times \mathbf{u}_n) = 2\omega_n \delta_{mn} \qquad (22)$$

is the necessity of truncating the spherical harmonic expansion of the Coriolis force term $(\mathbf{u}_m^*, \mathbf{\Omega} \times \mathbf{u}_n)$; this immediately leads to a quasi-orthogonalisation condition. In addition, in a rotating system, the notion of a "mode" implies the following summation of spherical harmonic terms

$$\mathbf{u}_n(\mathbf{r}) = \sum_{l=|m|}^{\infty} (\mathbf{S}_l^m + \mathbf{T}_l^m) \qquad (23)$$

cf. equation (16) and this must of course be truncated to implement a numerically feasible calculation. The formal solution to (21) for a unit step source is

$$\mathbf{S}(\mathbf{r}, t) = \sum_n Re\left\{(\mathbf{u}_n^*, \mathbf{f}_\infty)\frac{1}{\omega_n^2}(1 - e^{i\omega_n t})\mathbf{u}_n(\mathbf{r})\right\}, \quad t \geq 0 \qquad (24)$$

as given by Chao [1982]. For the numerical calculation using the above formulae, we have to be satisfied with a number of compromises which degrade the calculation. A hierarchy of schemes is sketched in Table 5, ranging from a "best" solution (which still ignores Earth ellipticity) to a "poor" solution which treats each spheroidal mode in the solid regions of the Earth independently and forces severe simplification of the Coriolis-coupled core eigenfunctions. I am forced to admit that somewhat below this "poor" solution would rate the very preliminary calculation presented here where the Earth is treated without rotation.

Conclusions

Despite the simplicity of the calculation presented here, we can perhaps venture some statements concerning core mode excitation by earthquakes. The small shell displacements and surface gravity perturbations render the excitation of a single mode to be well below current detection levels for a reasonably-sized earthquake such as considered by Melchior et al. [1988]. Even under very favourable conditions (i.e. a moment $M = 10^{30}$dyne cm. earthquake) it is not possible to exceed a 1 ngal signal at the surface. The amplitude of the spectral peaks in the Brussels record is up to 11 ngals with a background noise of about 4 ngals. Unless there are such events as "corequakes" of comparable energy release as earthquakes, (and it is impossible of course to have dislocations in a fluid) we need to proceded with caution in ascribing surface observations to core motions.

Possibly helping in favour of detection is the effect of summing terms of different spherical harmonic degrees, as indicated in equation (23). Ultimately, when a wide spectral response is computed (e.g. for a superconducting gravimeter), the effect of spreading the energy amongst many neighbouring modes in the core may not enhance the general level of excitation above the independent mode calculation presented here.

Acknowledgments. Support for this project is provided by an NSERC Operating Grant # A4240.

References

Alterman, Z, H. Jarosch and C.L. Pekeris, Oscillations of the Earth, *Proc. Roy. Soc. London, Ser. A*, *252*, 80-95, 1959.

Chao, B. F., Excitation of normal modes on non-rotating and rotating Earth models, *Geophys. J. Roy. astr. Soc.*, *68*, 295-315, 1982.

Crossley, D., Oscillatory flow in the liquid core, *Phys. Earth Planet. Int.*, *36*, 1-16, 1984.

Dahlen, F. A., The excitation of the Chandler wobble by earthquakes, *Geophys. J. Roy. astr. Soc.*, *25*, 157-206, 1971.

Dahlen, F. A, and M. L. Smith, The influence of rotation on the free oscillations of the Earth, *Phil. Trans. Roy. Soc. London, Ser. A*, *279*, 583-629, 1975.

Friedlander, S., Internal oscillations in the Earth's fluid core, *Geophys. J. Roy. astr. Soc.*, *80*, 345-361, 1985.

Gilbert, F., Excitation of the normal modes of the Earth by earthquake sources, *Geophys. J. Roy. astr. Soc.*, *22*, 223-236, 1970.

Jackson, B. V. and L. B. Slichter, The residual daily Earth tides at the South Pole, *J. Geophys. Res.*, *79*, 1711-1715, 1974.

Johnson, I. M. and D. E. Smylie, A variational approach to whole-Earth dynamics, *Geophys. J. Roy. astr. Soc.*, *50*, 35-54, 1977.

Jordan, T. H. and D. L. Anderson, Earth structure from free oscillations and travel times, *Geophys. J. Roy. astr. Soc.*, *36*, 411-459, 1974.

Kovach, R. L. and D. L. Anderson, Study of the energy of the free oscillations of the Earth, *J. Geophys. Res.*, *72*, 2155-2168, 1967.

Mansinha, L., D. E. Smylie and C. H. Chapman, Seismic excitation of the Chandler wobble revisited, *Geophys. J. Roy. astr. Soc.*, *59*, 1-17, 1979.

Masters, G., Observational constraints on the chemical and thermal structure of the Earth's deep interior, *Geophys. J. Roy. astr. Soc.*, *57*, 507-534, 1979.

Melchior, P. and B. Ducarme, Detection of inertial gravity oscillations in the Earth's core with a superconducting gravimeter at Brussels, *Phys. Earth Planet. Int.*, *42*, 129-134, 1986.

Melchior, P., D. J. Crossley, V. D. Dehant and B. Ducarme, Have inertial waves been identified from the Earth's core ?, *AGU*, (in press).

Pekeris, C. L. and Y. Accad, Dynamics of the liquid core of the Earth, *Phil. Trans. Roy. Soc. London, Ser. A*, *273*, 237-260, 1872.

Smith, M. L., Translational inner core oscillations of a rotating, slightly elliptical Earth, *J. Geophys. Res.*, *81*, 3055-3065, 1976.

Smylie, D. E. and L. Mansinha, The elasticity theory of dislocations in real Earth models and changes in the rotation of the Earth, *Geophys. J. Roy. astr. Soc.*, *23*, 329-354, 1971.

Smylie, D. E. and M. G. Rochester, Long period core dynamics, in *Earth Rotation : Solved and Unsolved Problems ; Proc. NATO Advanced Study Workshop*, 297-324, D. Reidel, 1986.

Stacey, F. D., B. J. Brennan and R. D. Irvine, Finite strain theories and comparisons with seismological data, *Geophys. Surv.*, *4*, 189-232, 1981.

Verhoogen, J., *Energetics of the Earth*, National Academy Press, 139pp., 1980.

Zürn, W., B. Richter, P. A. Rydelek and J. Neuberg, Comment on : "Detection of inertial gravity oscillations in the Earth's core with a superconducting gravimeter at Brussels", *Phys. Earth Planet. Int.*, *49*, 176-178, 1987.

INERTIAL WAVES IN A DIFFERENTIALLY ROTATING FLUID

Sterling Gunn[1] and Keith D. Aldridge

Center for Research in Experimental Space Science, York University,
North York, Ontario, Canada M3J 1P3

Abstract. A perturbation method is developed which will permit the calculation of inertial wave eigenfrequencies for a Poincaré model of the Earth's fluid outer core in which the rotation speed is non-uniform. The method is demonstrated by calculating the frequency for an axisymmetric wave in a cylindrical cavity using differential rotations obtained from a numerical model. The results indicate a sensitivity of the eigenfrequency shifts to the profile of the differential rotation. The relevance of the associated geophysical inverse problem is discussed.

Introduction

The identification of inertial waves in the Earth's liquid core by [Aldridge and Lumb, 1987] in the gravimetric data [Melchior and Ducarme, 1986] is the first step in the measurement of both thermodynamical and dynamical properties of the core. This paper is concerned with the development of a method to calculate the effect of a non-uniform rotation on inertial wave eigenfrequencies so that ultimately their measurement can be used to determine the departure of the fluid core from uniform rotation using inverse methods.

Theoretical discussions of inertial wave frequencies usually neglect non-linear phenomena, and assume the fluid rotates as a solid body. With these assumptions, it has been shown by Greenspan [1968] that for a contained fluid there exists a spectrum of frequencies whose magnitudes are bounded by zero and twice the magnitude of the solid body rotation rate, Ω. The distribution of eigenfrequencies within these bounds will be modified to account for a small differential rotation, and a specific example is discussed in the context of a liquid-filled cylindrical cavity for which a significant amount of theoretical and experimental work exists.

Early experimental work with the forced excitation of axisymmetric inertial waves in a spherical cavity [Aldridge, 1967] found that a mean flow and a change in the inertial wave frequencies were related to the amplitude of the forcing. The amplitude of the mean flow was found to be proportional to the square of the forcing amplitude. McEwan [1970] found mean azimuthal flows with magnitudes up to 10% of the solid body rotation rate. Aldridge [1967] found that the frequency for the peak resonant response shifted as the forcing amplitude increased. McEwan [1970] and Stergiopoulos and Aldridge [1982] (hereinafter SA) found that as the forcing amplitude increased, the resonance would disappear to be replaced by an irregular response as shown in Figure 5 of SA, and this was described as resonance collapse.

Formulation

The above results suggest a link between the presence of a mean flow, and changes in the inertial wave frequencies. To investigate these findings, it was assumed that the mean flow is described by a velocity, $\overline{u}$, and a pressure, P, satisfying the governing equations. Both were functions of the spatial variables, but independent of time. We do not consider the origin of the mean flow in the present work. It then remains to determine the dependence of the frequencies on the mean flow.

The pressure and velocity fields were assumed to have the following forms

$$\begin{aligned} p &= P + \epsilon p', \\ u &= \overline{u} + \epsilon u', \end{aligned} \tag{1}$$

where the primed quantities are small disturbances having a time dependence of the form $e^{i\lambda t}$, and ϵ is the size of the disturbance relative to the mean flow. The unknown velocity field was iteratively eliminated, to $\mathcal{O}(\epsilon)$,

[1]Department of Applied Mathematics, University of Western Ontario, London, Ontario, Canada N6A 5B8

from the governing equations to obtain an equation for the unknown pressure field:

$$\begin{aligned} \nabla^2 p \; - \; & 4\frac{(\hat{k}\cdot\nabla)^2 p}{\lambda^2} \qquad (2)\\ + \; & \frac{\epsilon\nabla\cdot}{\lambda(\lambda^2-4)}\{16i\,\frac{\hat{k}(\hat{k}\cdot\vec{L}(\hat{k}(\hat{k}\cdot\nabla)p))}{\lambda^2}\\ + \; & 8[\hat{k}(\hat{k}\cdot\vec{L}(\hat{k}\times\nabla p)) + \hat{k}\times\vec{L}(\hat{k}(\hat{k}\cdot\nabla)p]\\ - \; & 4i[\hat{k}(\hat{k}\cdot\vec{L}(\nabla p)) + \hat{k}\times\vec{L}(k\times\nabla p) + \vec{L}(\hat{k}(\hat{k}\cdot\nabla)p)]\\ - \; & 2\lambda[\hat{k}\times\vec{L}(\nabla p) + \vec{L}(\hat{k}\times\nabla p)]\\ + \; & i\lambda^2\vec{L}(\nabla p)\}\\ + \; & \mathcal{O}(\epsilon^2) = 0, \end{aligned}$$

where

$$\vec{L}(\vec{v}) = (\overline{u}\cdot\nabla)\vec{v} + (\vec{v}\cdot\nabla)\overline{u}.$$

The coefficients of this equation are dependent upon the mean velocity field.

The boundary conditions for the pressure field are obtained by requiring the normal component of the velocity to be continuous at all rigid boundaries.

The $\mathcal{O}(1)$ terms, when equated to zero and coupled with the boundary conditions given above, form the Poincaré problem. This problem produces an infinite number of eigenfunction-eigenvalue pairs, each associated with a particular inertial mode.

For a differential rotation, equation 2 admits a separable solution, of the form $p(r,\theta,z) = R(r)Z(z)e^{ik\theta}$, in a cylindrical coordinate system, with the z axis corresponding to $\hat{k}$. The z dependence, Z, of the pressure amplitude is independent of the frequency λ and the perturbation parameter ϵ. Thus the solution's dependence on ϵ is limited to the frequency and the radial dependence, R, of the pressure amplitude.

The problem determining R, λ is an eigenvalue problem containing the small parameter ϵ, and so a solution is sought by expanding both R and λ as power series in ϵ:

$$R(r) \sim R(r)_o + \epsilon R(r)_1 + \mathcal{O}(\epsilon^2),$$

$$\lambda \sim \lambda_o + \epsilon\lambda_1 + \mathcal{O}(\epsilon^2).$$

Substitution of these expressions into the eigenvalue equation for R leads to an equation of the form

$$(H_o + \epsilon H_1)(R_o + \epsilon R_1) + (a_o + \epsilon a_1)(R_o + \epsilon R_1) + \mathcal{O}(\epsilon^2) = 0, \qquad (3)$$

where

$$H_o = \frac{\partial}{r\partial r}\left(r\frac{\partial}{\partial r}\right) - \frac{k^2}{r^2} \quad ,$$

$$\begin{aligned} H_1 = \; & \frac{\Omega\Gamma + 2\lambda_o r\Omega'}{\lambda_o(\lambda_o^2-4)}\frac{\partial}{r\partial r}\left(r\frac{\partial}{\partial r}\right)\\ + \; & \frac{\Omega\Gamma + 2\lambda_o(6-k\lambda_o)r\Omega' + 2\lambda_o r^2\Omega''}{\lambda_o(\lambda_o^2-4)}\frac{\partial}{r\partial r}\\ - \; & \frac{k^2\Gamma\Omega + 2k(6+\lambda_o^2-3k\lambda)r\Omega' + 4kr^2\Omega''}{\lambda_o(\lambda_o^2-4)r^2}\\ + \; & \frac{k\Omega(\lambda_o^2-4)\gamma^2}{\lambda_o^3}\\ a_o = \; & -\lambda_o^2 - 4\lambda_o^2\gamma^2 \quad ,\\ a_1 = \; & -8\lambda_1\frac{\gamma^2}{\lambda_o^3} \quad ,\\ \Gamma = \; & 8\lambda_o - 4k - k\lambda_o^2 \quad . \end{aligned}$$

The coefficient of each power of epsilon must vanish independently, leading to a sequence of problems determining the unknown coefficients R_n, λ_n :

$$\mathcal{O}(1) \qquad H_oR_o + a_oR_o = 0$$

$$\mathcal{O}(\epsilon) \qquad H_oR_1 + H_1R_o + a_oR_1 + a_1R_o = 0$$

The first problem has Bessel functions as solutions,

$$J_{|k|}(\xi_{nmk}r)$$

with the corresponding eigenvalue ξ_{nmk} determining the mode's frequency [Greenspan, 1968]

$$\lambda_{o,nmk} = 2\left(1 + \frac{\xi_{nmk}^2}{n^2\pi^2}\right)^{-1/2} \quad .$$

The modal indices are: n, the axial wave number, k, the azimuthal wave number, and m, the radial wave number.

The second problem is solved utilizing the results of the first problem. The correction $R_{1,nmk}$ is expanded in terms of the basis $\{J_{o,nmk}\}$:

$$R_{1,nmk} = \sum_{n'm'k'} A_{n'm'k'}J_{o,n'm'k'} \qquad (4)$$

The case $n = n'$, $m = m'$, $k = k'$ leads to the determination of the frequency correction:

$$\lambda_{1,nmk} = -\frac{\lambda_{o,nmk}(R_{o,nmk}, H_{1,nmk}, R_{o,nmk})}{8n^2\pi^2(R_{o,nmk}, R_{o,nmk})}$$

where the indices on the operator H_1 indicate that $\lambda_{o,nmk}$ is to be used in evaluating the form of the operator, and (f,g) indicates an inner product over the radius of the cylinder. This correction depends only upon known quantities: the basis functions $\{J_{o,nmk}\}$, the unperturbed frequencies $\lambda_{o,nmk}$, and the operator $H_{1,nmk}$.

Results

The forced excitation of axisymmetric inertial waves in a liquid filled cylindrical cavity, with aspect ratio $\alpha = R/H = 0.9974$, has been numerically modeled at the Ballistic Research Laboratory. Data from this model were used to generate two differential rotations for the $(1,0,0)$ inertial mode, having a frequency of $\lambda = 1.267$. These models were the same except for the values of the half amplitude of the forcing, which were 0.09, and 0.18 radians. The latter value corresponds to a regime in which experiments have shown the response to be irregular. Increasing the amplitude not only increased the amplitude of mean flow, but also altered the profile of the mean flow. This suggests that the linear model, which would have a simple rescaling of the the mean flow as the forcing was increased, is inadequate.

Frequency corrections for the case $\lambda = 1.267$ are given in Table 1. The two-fold increase in forcing amplitude did not bring about a two-fold increase in the frequency correction, a result accredited to the change in the profile of the mean flow.

TABLE 1. Frequency Corrections for $\lambda_{100} = 1.267$

ϵ	$\epsilon\lambda_1$
0.09	0.019
0.18	0.018

Conclusions

The alteration of the inertial wave frequencies by a mean flow, coupled with the high Q of inertial waves, offers an explanation of resonance collapse. Initially, inertial waves are excited in a fluid rotating as a solid body. Non-linear interactions produce a non-uniform flow, altering the natural frequency of the inertial wave. The forcing frequency is not altered, and so the system is no longer in resonance, reducing the amplitude of the response.

It is a straightforward matter to cast the above analysis for a spherical geometry in order to make a direct geophysical application. Once inertial wave eigenfrequency corrections are available for mean flows in this geometry, it should be possible to use measured inertial wave frequencies to determine the mean flow structure. Implicit in this statement, of course, is the assumption that all other contributions to the eigenfrequency shifts, such as stratification, compressibility and self gravitation have been taken into account. In fact we anticipate that all of these contributors to the eigenfrequency magnitudes will be included simultaneously in any future core model. Ultimately the success of this method will be determined by our understanding of the processes which determine the mean flow and the accurate determination of inertial wave frequencies in the presence of such a flow.

References

Aldridge, K.D., *PhD thesis*, M.I.T., Cambridge, Mass., 1967.

Aldridge, K.D. and L.I. Lumb, Inertial waves identified in the Earth's fluid outer core, *Nature, 325*, 421-423, 1987.

Greenspan, H.P., *The Theory of Rotating Fluids,* Cambridge University Press, 328 pp., 1968.

McEwan, A.D., Inertial oscillations in a rotating fluid cylinder, *J. Fluid Mech., 40,* 603-640, 1970.

Melchior, P., and B. Ducarme, Detection of inertial gravity oscillations in the Earth's core with a superconducting gravimeter at Brussels, *Phys. Earth planet. Int., 42,* 129-134, 1986.

Stergiopoulos, S., and K.D. Aldridge, Intertial waves in a fluid partially filling a cylindrical cavity during spin-up from rest, *Geophys. Astrophys. Fluid Dyn., 21*, 89-112, 1982.

EVIDENCE FOR LATERAL HETEROGENEITY AT THE CORE-MANTLE BOUNDARY FROM THE SLOWNESS OF DIFFRACTED S PROFILES

Michael E. Wysession and Emile A. Okal

Department of Geological Sciences, Northwestern University, Evanston, Illinois 60208

Abstract. A regional study of SH velocities in D" has been conducted through the comparison of the apparent ray parameters of diffracted SH waves [*Sd*] from large earthquakes and synthetic pulses generated by normal mode summation for three different velocity structures. This technique has the advantage of automatically including all non-geometric optics effects in the calculation of the synthetics and of avoiding the complications due to varying waveform frequency content that arise when trying to directly convert apparent slownesses into mantle velocities.

We computed the excitation functions of torsional normal modes up to 0.05 Hz for the smooth model PREM, a model PREM(HVZ) incorporating Lay and Helmberger's [1983a] proposed high-velocity zone, and a PREM(LVZ) low-velocity structure. The apparent slownesses for the path profiles were obtained by cross-correlations of both the synthetic *Sd* arrivals, corrected for ellipticity using the modal approximations of Dahlen [1975], and the data, which were filtered and corrected for upper mantle heterogeneity (using the tomographic SH velocity models of Woodhouse and Dziewonski [1983], and Tanimoto [1987] so as to be compatible with the synthetics). Comparisons of the cross-correlated slownesses revealed one region on the CMB with D" velocities slower than those of PREM, one region compatible with PREM, two zones distinctly faster than PREM and one zone that strongly represents a region of high velocity. This high-velocity zone was also identified as such by the Lay and Helmberger [1983a] study.

Introduction

The structure and dynamics of the so-called D" region at the base of the mantle have continued over the past two decades to arouse much interest and controversy among a variety of disciplines. Most early models of the shear velocity above the core-mantle boundary (CMB) displayed a homogeneous low velocity zone [Cleary, 1969; Hales and Roberts, 1970; Bolt et al., 1970], which is consistent with the concept of D" as a simple thermal boundary layer [Jeanloz and Richter, 1979]; later models suggested heterogeneous regions of low velocity [Bolt and Niazi, 1984]. However, subsequent studies have presented a wider variety of models, some with high velocity zones in D" [Mitchell and Helmberger, 1973; Mula and Müller, 1980; Lay and Helmberger, 1983], and recent work by several authors has found evidence for large scale heterogeneity at the base of the mantle [Gudmundsson et al., 1986; Woodhouse and Dziewonski, 1987]. The determination of the seismic velocities in D" is of great importance in discerning the dynamic processes at work along the CMB. The average values of the seismic velocities should provide insight into the general thermal and mineralogical structure of D", while lateral heterogeneity may help map areas of upward convection (expected to feature low velocities and low-Q) and, conversely, zones of sinking of colder convected mantle (expected to be faster), as suggested, for example, by Olson et al. [1987].

While studies of D" have utilized a wide range of techniques, we concentrate in this paper on the determination of the apparent slowness of diffracted S [hereafter *Sd*] arrivals (Figure 1). We are motivated to a large extent by the results of Lay and Helmberger [1983a], who interpreted teleseismic S arrivals in the distance range 75—90° as triplications due to a zone of high S velocities approximately 280 km thick and 2.75% fast. If confirmed, such a structure would have important implications for the structure and mineralogy of D". These authors' interpretation was however non-unique, since the later arrivals making up the triplication could be due to near-source heterogeneity or the effect of scattering along the CMB. A summary of these arguments and of the ensuing controversy can be found in Schlittenhardt et al. [1985], Haddon and Buchbinder [1986] and Lay [1986].

Difficulties with *Sd* Ray Parameters

The use of a profile of diffracted S waves leaving the source at a common azimuth should in principle alleviate this problem, since the downswing part of the path from the source to the core is common to all rays in the profile, and near-source complexity cancels out of the apparent slowness. We refer to Okal and Geller [1979], Mula and Müller [1980], and Bolt and Niazi [1984] for previous work on *Sd*; in general, these studies successfully measured the slowness $p = dT/d\Delta$ of the wave along a profile of stations. As pointed out for example by Aki and Richards [1980], the interpretation of p in terms of the shear-wave velocity at the CMB is not simple because the diffracted arrivals, in violation of the laws of geometrical optics, propagate along the CMB with varying frequencies travelling through varying depths into D", the longer wavelengths of the lower frequencies sampling further up into the mantle. This causes complications in several manners: (1) for a D" model featuring a velocity gradient, dispersion of the diffracted ray will occur with different frequencies travelling through regions of different velocity; (2) as the ray propagates along the CMB, its amplitude decreases (due to diffraction back into the mantle) as $\exp[-a\,\omega^{1/3}\Delta]$, where Δ is distance along the CMB and ω angular frequency [Aki and Richards, 1980, p. 440]; (3) anelastic attenuation further reduces the high frequency components of the signal.

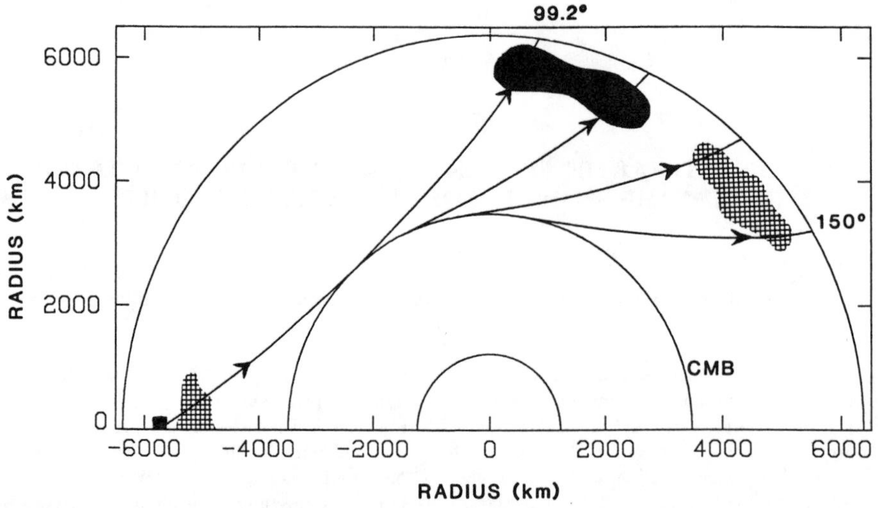

Fig. 1. Sketch of the path of four diffracted S (*Sd*) waves from a 600 km deep earthquake, generated for a PREM earth [*Dziewonski and Anderson,* 1981]. On a geometric basis the first *Sd* arrivals would appear after 99.2°, and the phase is found beyond 150°. The downswing paths are not differentiated by mantle heterogeneity, which is not the case for the *Sd* upswing paths.

As a result, each arrival with a different epicentral distance will have a different waveform due to its varying frequency content [Chapman and Phinney, 1972]. This is particularly important because it means that the value of the slowness obtained from cross-correlation along a particular path will be dependent upon the relative locations of the stations, and one would expect the slowness along a given portion of the CMB to vary with the addition and removal of stations. Mula and Müller [1980] recognized this problem and found that the apparent ray parameter varied with different dominant frequencies for synthetic seismograms computed by the reflectivity method. The value of the slowness will also depend upon the technique used in measuring the trend, and all methods will be arbitrary to some degree as they are subject to the varying frequency content of the arrivals.

For these reasons, in the present paper, we retrieve only the slowness from the data, and use forward modeling to determine an appropriate D" velocity structure by generating synthetic *Sd* waves from a given initial velocity model and comparing the synthetics' slowness with that of the data. An initial velocity model is deemed appropriate only to the extent of the similarity between the two slownesses. The synthetics are computed by normal mode theory, which has the advantage of including all diffraction effects automatically [Okal, 1978].

While the ray parameter from a set of *Sd* waves is, to the first order, the linear trend of the successive arrivals at increasing distances along the CMB, there are several factors which must be corrected or accounted for. It is very important, for example, that a narrow azimuthal range be accepted in the inclusion of arrivals for a given path. Recent work by Cormier [1987] has shown that contamination of *S* rays by slab-diffracted energy can cause a difference in arrival times of several seconds for rays leaving an event at different azimuths. In addition, rays at varying azimuths could accrue travel-time differences due to their different downswing paths through the heterogeneous mantle. A narrow azimuthal range not only eliminates errors due to source effects and downswing heterogeneity but also allows the rays to sample the same portions of the CMB.

A problem which has not been addressed in previous studies of diffracted ray slownesses has been the effect of ellipticity. This effect, though usually not large, is variable depending upon the azimuth of a profile and the distribution of stations along it. For example, in a set of *Sd* arrivals starting at high latitudes and approaching the equator, successive arrivals would travel increasingly greater distances through the mantle, causing a bias toward a larger ray parameter and therefore erroneously implying a D" low velocity zone. We have corrected for the ellipticity by including modal approximation terms in the generation of our synthetic seismograms.

Finally, a problem that previous *Sd* studies have not been able to correct for is the differing *Sd* upswing paths, seen in Figure 1. Diffracted S waves can arrive over a wide distance range, often causing the upswing paths to rise through different regions of the notoriously heterogeneous upper mantle, even for rays leaving a single event with the same azimuth. If, for example, the closest *Sd* arrivals are through a fast continental shield region and the furthest rays come up through a slow young oceanic structure, the trend through the arrivals would be biased towards a larger ray parameter value and again would falsely imply a D" low velocity zone. We address this problem by using available models of upper mantle lateral heterogeneity to compute travel-time corrections over all upswing paths, before computing the experimental slowness for each profile. The resulting corrections are significant along certain profiles.

Data and Analysis

A total of 71 *Sd* arrivals were used from 38 WWSSN and Canadian (and the Palisades Press-Ewing) stations, allowing the determination of ray parameters for 12 paths along 5 different regions of the CMB. Both horizontal components were digitized, interpolated at 0.5 s intervals using a cubic polynomial scheme and rotated into their transverse components. In the PREM model [Dziewonski and Anderson, 1981] the calculated first *Sd* rays arrive at epicentral distances of $\Delta = 99.2°$ for an event at 600 km

TABLE 1. Earthquakes Used in This Study

No.	Region	Date	Origin Time	Depth (km)	Epicenter °N	Epicenter °E	m_b	Focal Mechanism ϕ	δ	λ	Ref.
1	New Guinea	4/24/64	05:56:09.8	99	-5.07	144.20	6.3	183	80	-90	a
2	Loyalty Is.	10/7/66	15:55:11.3	165	-21.59	170.56	6.0	107	70	21	a
3	Tonga	10/9/67	17:21:46.0	605	-21.10	-179.20	6.2	54	85	-83	a
4	Tonga	2/10/69	22:58:03.3	635	-22.75	178.76	6.0	96	86	-65	b
5	Burma	7/29/70	10:16:20.4	68	26.02	95.37	6.4	197	60	165	c
6	Kuril Is.	1/29/71	21:58:05.4	515	51.69	150.97	6.0	35	71	-90	d
7	Tonga	11/20/71	07:27:59.5	533	-23.45	-179.88	6.0	160	75	100	d
8	Tonga	2/22/75	22:04:33.5	333	-24.98	-178.88	6.1	35	87	-143	d
9	Honshu	6/29/75	10:37:40.6	549	38.79	130.09	6.1	222	78	120	d
10	Honshu	3/9/77	14:27:53.6	579	41.61	130.88	5.9	194	75	76	e

References for focal mechanisms: a= *Isacks and Molnar* [1971]; b= *Mula* [1981]; c= *Tandon and Srivastava* [1975]; d= This study; e= *Giardini* [1984].

depth, and $\Delta = 101.6°$ for a surface event, so we tried to avoid using any arrivals closer than 100°, only twice using rays that could contain mostly undiffracted S energy. The SH energy decays quickly along the CMB, but we found as appropriate for the study four of the Sd arrivals recorded beyond 150°, making the range of the CMB sampled by a given profile to be as large as 50°, or approximately 3000 km. It was always our attempt to get as many arrivals with as wide a distance range as possible for each path used. While there is most probably inhomogeneity in D" on a scale smaller than 3000 km, and a slowness retrieved over this range will be a large scale average, this trade-off was propitious due to the more precise determination of the ray parameter using a broader distance range. Tables 1 and 2 list epicentral and path information for the records used in this study.

In order to retain large amounts of SH energy around the CMB, the events used in this study, all with $m_b > 5.9$, involved predominantly dip-slip motion as can be seen from the focal mechanisms listed in Table 1. The source parameters were either taken from previous studies or calculated on the basis of first arrivals that were both read directly from the data and taken from ISC bulletins. We tried to use primarily deep events for several reasons. First, the LVZ at the base of the lithosphere greatly attenuates S waves that pass through it, and rays from deeper events pass through this zone once instead of twice. The two earthquakes we used that were shallower than 100 km also happened to have the two greatest body magnitudes (m_b = 6.3 and 6.4). Records from deep events are usually cleaner and less noisy, making the Sd arrivals more distinct, but more importantly with deep events Sd is also distinct from the surface reflected sSd, simplifying the procedure to determine the slowness. Only in the case of Event 5, were Sd and sSd too close to be separated in the cross-correlation process, the consequences of which are discussed later.

Another requirement for the earthquake sources was that their SH radiation patterns be favorable for the azimuths along which a string of stations were located. The allowed range in azimuths from a given event to individual stations was usually limited to 10°, so the geographical locations of existing stations and large seismic events not only limited the CMB paths we could study but also forced us to use events with large relative radiation amplitudes. These can be calculated following Kanamori and Stewart [1976] as $R^{SH} = -q_L \cos i_h - p_L \sin i_h$, where q_L and p_L are the

TABLE 2. Characteristics of the Sd Profiles Used in This Study

No.	Ave. Azimuth (°)	R^{SH}	Stations used
1	335.5	0.831	NUR, KON, ESK, VAL, PTO
2	295.5	0.481	NDI, LAH, QUE, MSH, SHI, JER
3a	51.5	0.920	FLO, SCP, PAL, WES, HAL, STJ, PDA
3b	292.0	0.458	SHL, NDI, QUE, SHI
4a	54.0	0.566	OXF, AAM, BLA, SCB, SCP, GEO, OGD, STJ
4b	294.0	0.593	NDI, KBL, MSH, SHI, TAB, JER, HLW
5	0.0	0.230	AAM, SCP, FLO, BLA, OXF, ATL, SHA
6	55.5	0.754	BOG, QUI, NNA, ARE, LPB
7	291.0	0.589	NDI, QUE, SHI, JER, EIL
8	54.0	0.288	OXF, AAM, BLA, SCP, WES, STJ
9	35.5	0.506	ATL, SHA, BHP, BOG, QUI, LPB
10	32.5	0.712	ATL, SHA, CAR, BOG, LPB

radiation-pattern coefficients of Kanamori and Cipar [1974], and i_h is the take-off angle at the focal sphere, which here ranged from $i_h = 19.9°$ (at 68 km depth) to $i_h = 27.6°$ (at 635 km depth). With the optimal relative amplitude as $R^{SH} = 1$, at the azimuths of our profiles all of the events had values of $R^{SH} > 0.4$ with the exception of Events 5 ($R^{SH} = 0.23$) and 8 ($R^{SH} = 0.29$), which were retained because of their clear arrivals.

Finally, the data profiles were filtered with a low-pass filter of 0.05 Hz to make them compatible with the synthetic profiles. Figure 2 shows an example of *Sd* profile in the case of Event 4.

Upper Mantle Corrections

In order to further equate the filtered data with the synthetics, corrections for upper mantle heterogeneity along the *Sd* upswings were added to the travel times based on the tomographic full-mantle shear velocity model of Woodhouse [personal communication, 1987]. The upper mantle model of Tanimoto [1987] was also used, without significant differences.

The corrections were simply computed as

$$\Delta t = -\int_{\text{upswing}} \frac{\Delta\beta}{\beta^2}\, ds \qquad (1)$$

where the integral is taken along the upswing portion of the ray, and the lateral heterogeneity in shear velocity, $\Delta\beta$, is computed locally by summing its spherical harmonics components.

The range in arrival corrections to our data was 2.41 s ($\Delta t = -1.83$ s for Event 4 to SCB; $\Delta t = 0.58$ s for Event 2 to HLW). The greatest range in travel-time corrections for a single profile was 1.53 s (Event 4 to N. America), and the resulting slownesses reflect these biases.

Synthetic Seismograms and Ellipticity Corrections

Since the D" velocity structure cannot be directly inferred from the data ray parameters, synthetic seismograms for three different velocity models were used as a standard of comparison. The synthetics were generated using a summation of toroidal normal modes of free earth oscillations with the algorithm of Kanamori and Cipar [1974], and corrected for ellipticity using the modal approximations of Dahlen [1975]. This method has the advantage of being able to replicate diffracted SH arrivals, which as non-geometrical rays are not well modeled by some other techniques. For each velocity model the excitation functions with depth of almost 7000 modes (with frequencies up to 0.05 Hz and phase velocities above 8.0 km/s) were generated, and each synthetic seismogram was obtained by the summation of the modes; computational details can be found in Okal [1978]. The synthetics for each event were generated using the appropriate focal mechanisms listed in Table 1. As can be seen in Figure 2, using such a broad range of modes generates not only the diffracted *S* rays, but a complete set of SH body wave arrivals.

Velocity Models

The three velocity models used in creating the synthetics typify the types of models that have been invoked by many studies of D"; they are shown on Figure 3. All three of our models have a PREM structure down to a radius of 4075 km and have a CMB at $r = 3480$ km, but while one maintains the PREM values down to the CMB, the other two are perturbed to include high-velocity and low-velocity zones at the base of the mantle. The undisturbed PREM model, based on free earth oscillation data, is used to represent models having zero or small velocity gradients in D", actually including a very small negative gradient of $-8.42 \times 10^{-5}\ s^{-1}$ in the bottom 70 km and reaching a shear velocity of 7.2254 km/s at the CMB.

The HVZ model is in the manner of Lay and Helmberger [1983a], where evidence for a high-velocity zone was inferred from D" triplication patterns, and consists of 280 km of faster material reaching a maximum of 7.281 km/s at the CMB. Consistent with its initial design, this HVZ is discontinuously separated from a

TABLE 3. Cross-correlated Slownesses (s/deg)

Path Number	Region Sampled	Observed slowness: Raw Data ($T > 20$ s)	Observed slowness: Corrected (*Woodhouse*)	Synthetic slowness (*With Ellipticity*): PREM	HVZ	LVZ
2	Loyalty I. to Mid-East	8.51	8.47	8.55	8.42	8.73
3b	Tonga to Mid-East	8.50	8.47	8.56	8.40	8.75
4b	Tonga to Mid-East	8.46	8.46	8.55	8.44	8.74
7	Tonga to Mid-East	8.50	8.46	8.58	8.45	8.71
3a	Tonga to N.America	8.46	8.43	8.54	8.41	8.75
4a	Tonga to N.America	8.34	8.30	8.53	8.38	8.71
8	Tonga to N.America	8.46	8.42	8.55	8.38	8.73
6	Kuril Tr. to Americas	8.34	8.33	8.54	8.41	8.73
9	Honshu to Americas	8.33	8.30	8.57	8.43	8.76
10	Honshu to Americas	8.31	8.31	8.58	8.42	8.77
5	Burma to N.America	8.56	8.50	8.52	8.36	8.68
1	New Guinea to Europe	8.61	8.59	8.52	8.42	8.73

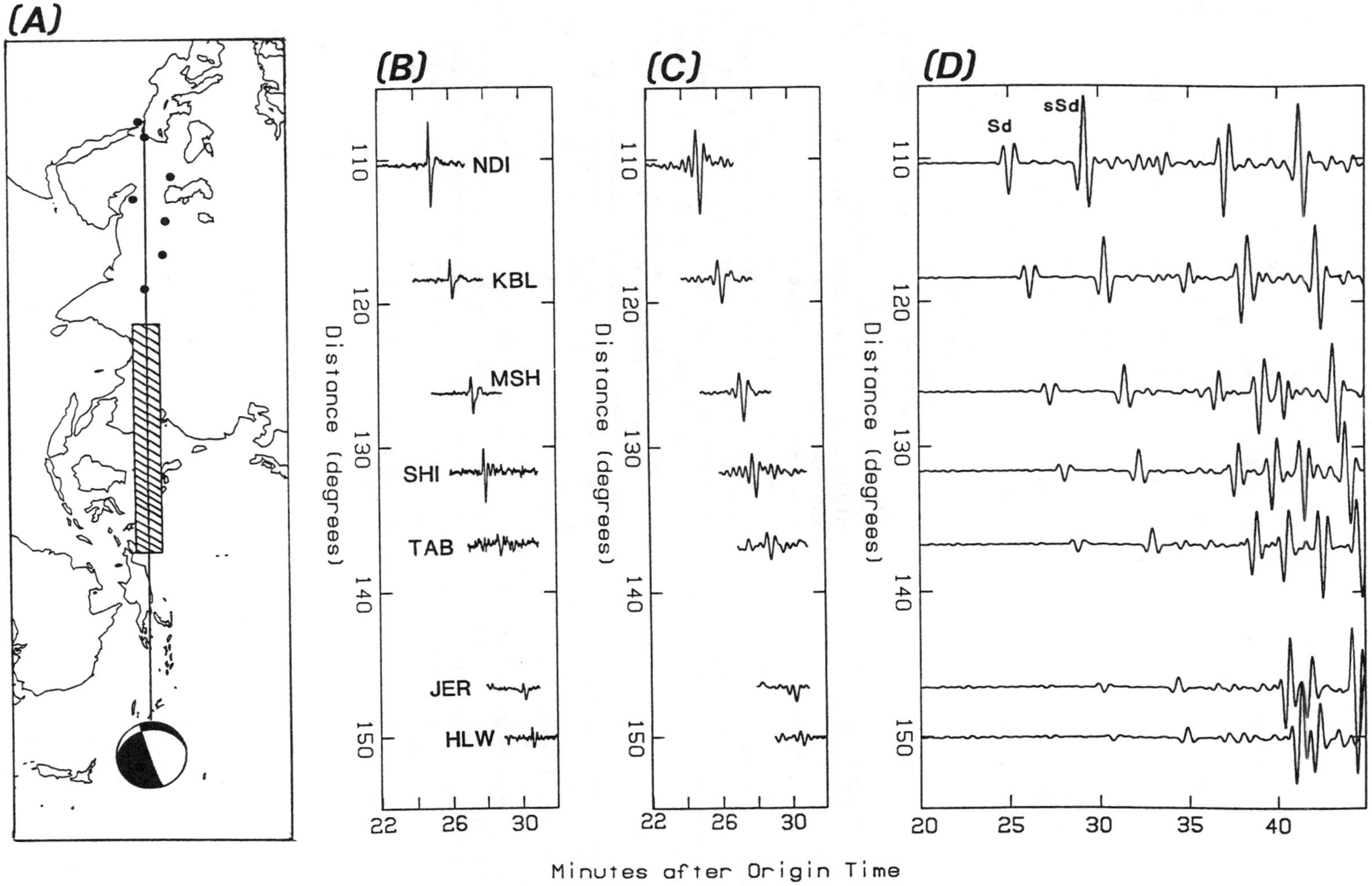

Fig. 2. Profiles for Event 4b to Asian and Mid-Eastern stations. (A) shows the path to the stations in relation to the focal mechanism and the region of CMB sampled (hatched). (B) and (C) show the *Sd* data before and after low-pass filtering at a cut-off frequency of 0.05 Hz. Note the amount of high-frequency energy still present out at 150°, in concurrence with observations of past studies [*Bolt et al.*, 1970]. (D) displays the PREM synthetic SH records from normal mode summation. In each case the apparent ray parameter corresponds to the linear trend through the *Sd* pulses.

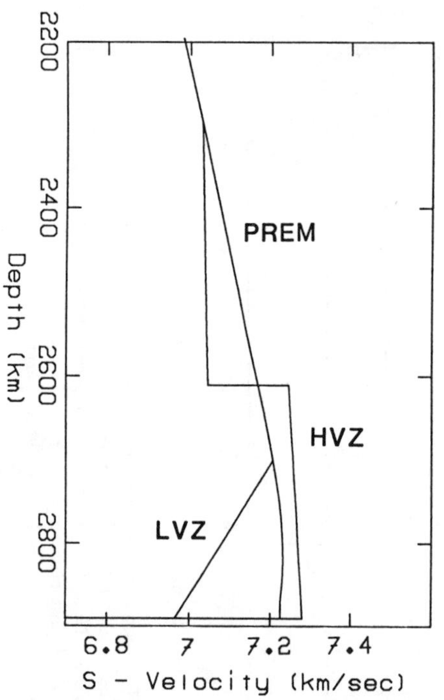

Fig. 3. The three D" SH-velocity structures used in the calculation of normal mode excitation functions for the generation of synthetic seismograms. The high-velocity model (HVZ) is in the style of the SLHO structure of *Lay and Helmberger* [1983*a*], and the low-velocity zone is modeled after M2 of *Mula and Müller* [1980].

zone of lower velocity that lies above it, so that the vertical travel-time of *ScS* will remain unchanged.

The LVZ, styled after model M2 of Mula and Müller [1980], is typical of low-velocity models in that its shear velocity reaches a maximum at the top of D" and decreases linearly with depth down to the CMB. At a radius of 3671 km the shear velocity reaches its maximum with $\beta = 7.2061$ km/s, and it decreases with a gradient of -1.26×10^{-3} s^{-1} down to a value of 6.965 km/s at the CMB. This is a modest relative (similar in structure but with a less pronounced negative velocity gradient) of several LVZ models previously presented, such as ANU2 of Hales and Roberts [1970] and that of Bolt [1972].

Cross-Correlation and Observations

The slownesses of both the *Sd* data and synthetics were determined using the cross-correlation method used by Okal and Geller [1979]. Since the data were filtered above 0.05 Hz and corrected for upper mantle heterogeneity, a comparison between the apparent ray parameters of data and synthetics is an appropriate means of testing the validity of our velocity models. The cross-correlation method, finding the best-fit linear trend between the *Sd* arrivals along a given path, involves isolating the *Sd* pulses, normalizing the signal energy of each pulse to unity and computing the cross-correlation functions $y_i \oplus y_j$ between the signals. The best-fitting slowness, p, corresponds to a maximum in the cross-correlation coefficient, F, determined by

$$F = \sum_i \sum_j y_i(t - p \cdot \Delta_i) \oplus y_j(t - p \cdot \Delta_j) \qquad (2)$$

Figure 4 shows the cross-correlation function $F(p)$ for the profile shown on Figure 2, and for the corresponding synthetics. Those paths which did not yield sharp and unitary peaks in the function $F(p)$ were discarded.

Previous studies of *Sd* waves have used visual picking methods in the determination of the slowness, choosing either the onsets [Bolt and Niazi, 1984] or peak deflections [Mula and Müller, 1980] of the arrivals, and had no means of correcting for the wave-forms changing due to the rapid high frequency decay. These methods would be very inappropriate for our study, with an absence of energy for periods $T < 20$ s in our pulses, but while the cross-correlation technique is also susceptible to bias from wave-form distortion, this distortion will appear in both data and synthetics and is therefore not problematic.

In all cases except Event 5, a tapered window was applied to isolate the *Sd* pulse before cross-correlation. For Event 5 (the Burma event and the shallowest earthquake used) the surface reflected *sSd* arrivals were so close to the *Sd* pulses that, with the absence of high frequencies, the two were distinct but inseparable. Therefore, the cross-correlated windows in these cases contained both signals, but given similar azimuths and take-off angles, the relative amplitudes and differential travel times between *Sd* and *sSd* arrivals should be the same at all stations and so should cause no bias in the calculated slowness.

Table 3 shows the slownesses for the paths used in our study. The first column is for the filtered data and the next includes the upper mantle corrections. It can clearly be seen that the addition of the heterogeneity corrections changes the slownesses by up to 0.06 s/°. In the tomographic model of Woodhouse [personal communication, 1987] we do see strong upper mantle velocity anomalies due to continent vs. ocean differences. Since all our stations are in continental regimes, we see a strong bias toward negative (fast) travel-time anomalies in our arrivals, but would have seen a greater variation in anomalies if we had used data from oceanic island stations, which tend to show a positive bias in travel times.

The last three columns of Table 3 show the synthetic slownesses from the three velocity models, with the ellipticity correction included. The effect of ellipticity varies depending upon the orientations of the profile path. For the two Honshu events this correction averaged +0.03 s/° among the three models, but was negligible for the Tonga profiles.

The variation in ray parameter due to the variable wave form frequency content from the unique juxtapositions of stations can be seen in the spherical model synthetics, which would otherwise yield identical slownesses for a given velocity model. For spherical PREM, which has the least vertical variance in D", the range in synthetic slownesses is 0.06 s/°, but for the other two models this range is 0.09 s/°, possibly due to their more complex structures. Add to this the corrections just mentioned and it is clear that in examining the path slownesses it is important not to attach significance to the individual values for the path-corrected data, but rather to their differentiation from the synthetic slownesses.

Figure 5 shows these comparisons as a function of the geographic region sampled at the CMB. "Average" shading is for slownesses between those of the PREM and HVZ models; "slow" implies velocities slower than PREM; "fast" implies velocities faster than our HVZ model. The four Tonga to Mid-East paths, sampling the CMB under the Western Pacific and Southeast Asia, all imply velocities close to the HVZ model but still possibly con-

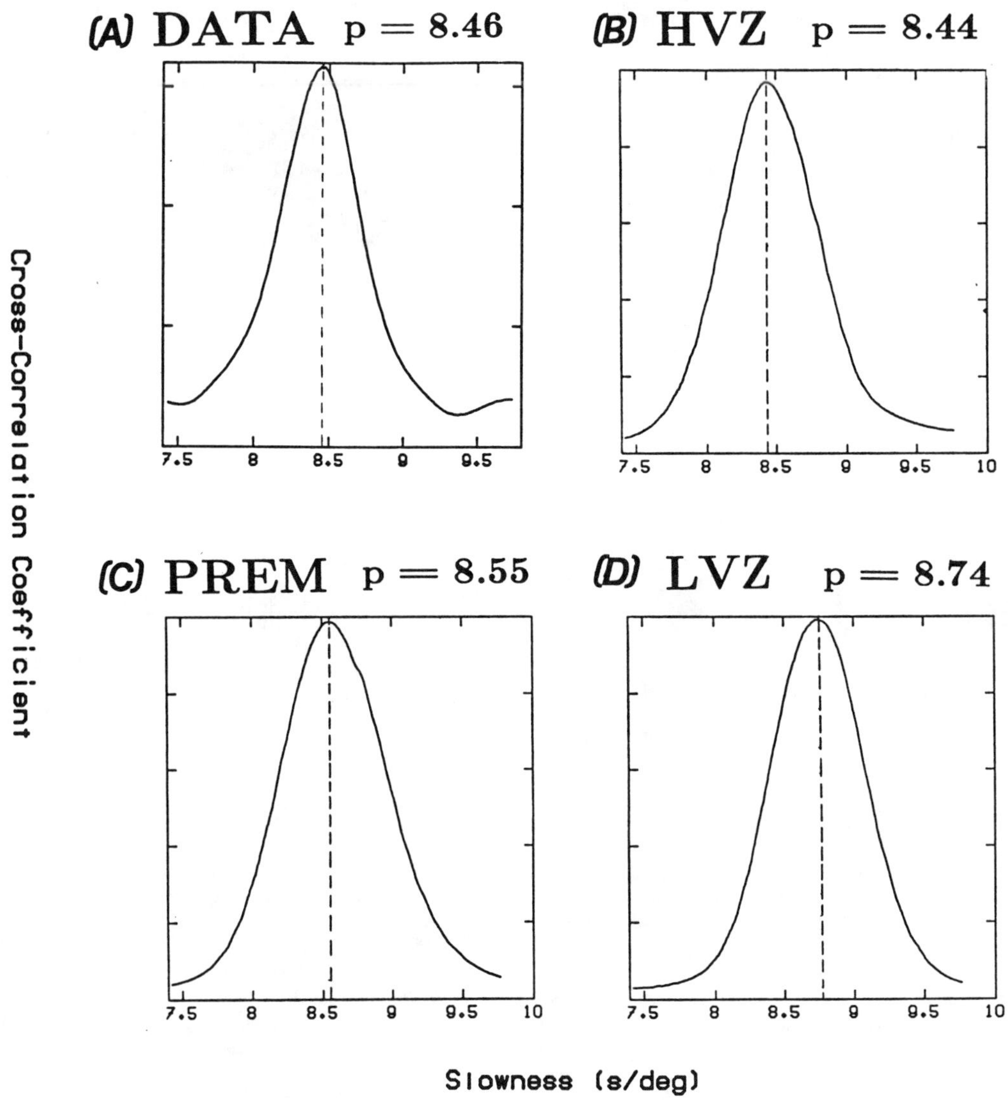

Fig. 4. Plots of the cross-correlation functions for the Event 4b profile. In (A), the peak represents the best-fitting apparent ray parameter through the filtered data arrivals (not corrected for upper mantle heterogeneity), and the following plots show the apparent slownesses through the synthetic *Sd* phases.

sistent with PREM. The three Tonga to North America paths suggest velocities even closer to our HVZ structure. The slowness for the Burma to North America path is consistent with that of the PREM synthetics.

All three Japan to America paths, however, show strikingly fast velocities, even faster than for our HVZ model. It is interesting to compare the geography of these results to that of Lay and Helmberger [1983b]. Because these authors' study is conducted at shorter distances, it is not possible to duplicate exactly their sampling. Nevertheless, their proposed area of high-velocity at the CMB, along the path from the Sea of Okhotsk to North America, falls in the immediate vicinity of our zone of high velocity (from Honshu and the Kuriles) to North and South America. Our HVZ was based on theirs in its structure but was also designed as a perturbation of PREM, maintaining *ScS* travel times. The Lay and Helmberger SLHO high-velocity zone had D" velocities in excess of 7.32 km/s (ours reached 7.28 km/s) and so would be more appropriate for our results than the HVZ we tested here. We do not have comparable coverage of their other region of high velocities (Argentina to North America).

The presence of a low-velocity zone was seen in only one region, the CMB under Siberia from our New Guinea to Europe profile. Even here, however, the slowness was only in between our PREM and LVZ models. Unfortunately, due to the paucity of stations in the Southern Hemisphere, we have no coverage of the CMB below the equator.

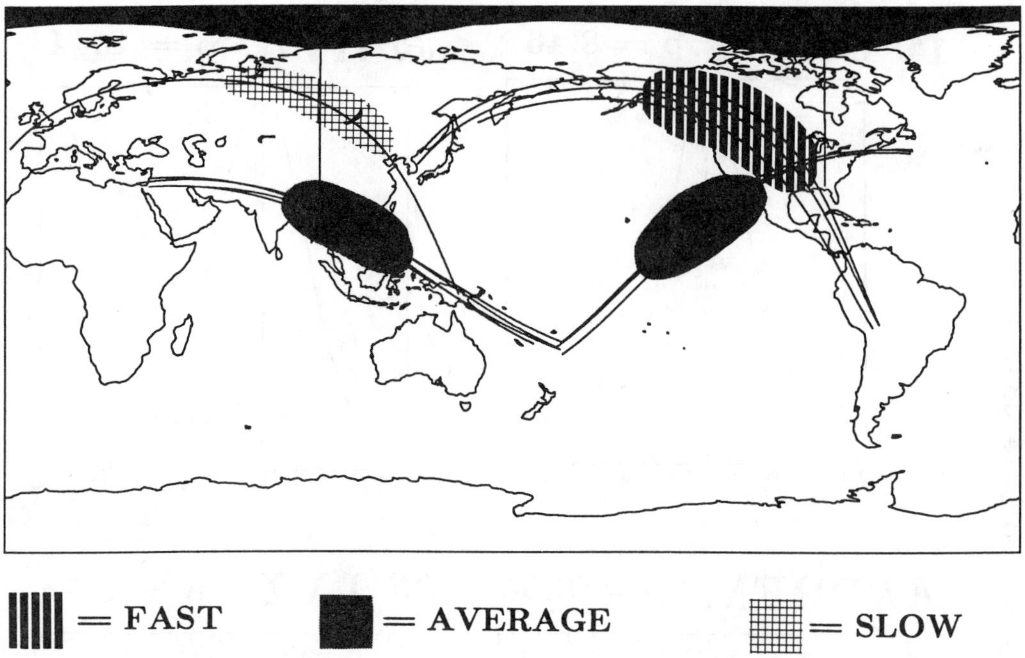

Fig. 5. Velocities in D" relative to the average of our data, which is slightly faster than for PREM. Solid lines represent average source-to-station great circle paths for the data profiles.

Conclusions

We find strong evidence for lateral heterogeneity in the velocity structure of the D" region at the base of the mantle. The use of diffracted SH waves is a powerful tool in examining this region, but considerations due to event depth and source parameters, station azimuth and location, ellipticity and slowness-determination techniques must first be taken into account. Both the ellipticity and upper mantle heterogeneity corrections can be significant for particular path profiles, and should be incorporated into future *Sd* ray parameter studies. In addition, inherent biases in the apparent ray parameter prevent a direct conversion into mantle velocities, but through comparison with synthetic seismograms generated through normal mode summation, we have been able to identify the appropriateness of different shear velocity models at different regions along the CMB. We find one region in D" that would be adequately modeled with a low-velocity perturbation to PREM (though not of the magnitude proposed by many other LVZ models), three regions that would be well represented by high-velocity perturbations to PREM, and one, in agreement with the results of Lay and Helmberger [1983b], that implies the presence of a very pronounced high-velocity zone. If high-velocity zones in D" were the result of the sinking of colder convected mantle, in the manner of Olson et al. [1987], and if full-mantle convection occurred, then the CMB directly below the Northern Pacific Ocean trenches, where oceanic lithosphere has continued to be subducted at a fast rate, would be a logical location.

Acknowledgements. We thank John Woodhouse for giving us a copy of his unpublished full-mantle 3-D shear wave velocity model, and Margie Yamasaki and the Lamont-Doherty Observatory for use of the seismology film chip library. This research was supported by NSF under grant EAR-84-5040.

References

Aki, K., and P. G. Richards, *Quantitative Seismology*, W. H. Freeman, San Francisco, 932 pp., 1980.

Bolt, B. A., The density distribution near the base of the mantle and near the Earth's center, *Phys. Earth Planet. Int., 5,* 301-311, 1972.

Bolt, B. A., N. Mansour, and M. R. Somerville, Diffracted ScS and shear velocity at the core boundary, *Geophys. J. R. astr. Soc., 19,* 299-305, 1970.

Bolt, B. A., and M. Niazi, S velocities in D" from diffracted SH-waves at the core boundary, *Geophys. J. R. astron. Soc., 79,* 825-834, 1984.

Chapman, C. H., and R. A. Phinney, Diffracted seismic signals and their numerical resolution, *Meth. Comput. Phys., 12,* 165-230, 1972.

Clayton, R. W., and R. P. Comer, A tomographic analysis of mantle heterogeneities from body wave travel time data, *Eos, Trans. Amer. Geophys. Un., 64,* 776, 1983 [abstract].

Cleary, J. R., The S velocity at the core-mantle boundary from observations of diffracted S, *Bull. seism. Soc. Am., 59,* 1399-1405, 1969.

Cleary, J. R., The D" region, *Phys. Earth Planet. Int., 30,* 13-27, 1974.

Cormier, V. F., Slab diffracted S waves, *Eos, Trans. Amer. Geophys. Un., 68,* 16, 1987 [abstract].

Dahlen, F. A., The normal modes of a rotating, elliptical Earth, *Geophys. J. R. Astron. Soc., 16,* 329-367,1968.

Dahlen, F. A., The correction of great circular surface wave phase velocity measurements for the rotation and ellipticity of the Earth, *J. Geophys. Res., 80,* 4895-4903, 1975.

Dahlen, F. A., and R. V. Sailor, Rotational and elliptical splitting of the free oscillations of the Earth, *Geophys. J. R. Astr. Soc., 58,* 609-623, 1979.

Doornbos, D. J., and J. C. Mondt, Attenuation of *P* and *S* waves diffracted around the core, *Geophys. J. R. astr. Soc., 57,* 353-379, 1979.

Doornbos, D. J., and J. C. Mondt, The interaction of elastic waves with a solid-liquid interface, with applications to the core-mantle boundary, *Pageoph, 118,* 1293-1309, 1980.

Dziewonski, A. M., and D. L. Anderson, Preliminary reference Earth model, *Phys. Earth Planet. Int., 25,* 297-356, 1981.

Giardini, D., Systematic analysis of deep seismicity: 200 centroid-moment tensor solutions for earthquakes between 1977 and 1980, *Geophys. J. R. astron. Soc., 77,* 883-911, 1984.

Gudmundsson, O., R. W. Clayton, and D. L. Anderson, CMB topography from ISC *PcP* travel times, *Eos, Trans. Amer. Geophys. Un., 67,* 1100, 1987 [abstract].

Haddon, R. A. W., and G. R. Buchbinder, Wave propagation effects and the Earth's structure in the lower mantle, *Geophys. Res. Letts., 13,* 1489-1492, 1986.

Hales, A. L., and J. L. Roberts, Shear velocities in the lower mantle and the radius of the core, *Bull. Seismol. Soc. Am., 60,* 1427-1436, 1970.

Isacks, B., and P. Molnar, Distribution of stresses in the descending lithosphere from a global survey of focal mechanism solutions of mantle earthquakes, *Rev. Geophys. Space Phys., 9,* 103-174, 1971.

Jeanloz, R., and F. M. Richter, Convection, composition, and the thermal state of the lower mantle, *J. Geophys. Res., 84,* 5497-5504, 1979.

Jordan, T. H., and K. C. Creager, Chemical boundary layers of the mantle and core, *Eos, Trans. Amer. Geophys. Un., 68,* 1494-1495, 1987 [abstract].

Kanamori, H., and J. J. Cipar, Focal process of the great Chilean earthquake, *Phys. Earth Planet. Int., 9,* 128-136, 1974.

Kanamori, H., and G. S. Stewart, Mode of the strain release along the Gibbs Fracture Zone, Mid-Atlantic Ridge, *Phys. Earth Planet. Int., 11,* 312-332, 1976.

Lay, T., Evidence of a lower mantle shear velocity discontinuity in *S* and *sS* phases, *Geophys. Res. Letts., 13,* 1493-1496, 1986.

Lay, T., and D. V. Helmberger, A lower mantle S-wave triplication and the velocity structure of D", *Geophys. J. R. astr. Soc., 75,* 799-837, 1983*a*.

Lay, T., and D. V. Helmberger, The shear-wave velocity gradient at the base of the mantle, *J. Geophys. Res., 88,* 8160-8170, 1983*b*.

Mitchell, B. J., and D. V. Helmberger, Shear velocities at the base of the mantle from observations of S and ScS, *J. Geophys. Res., 78,* 6009-6027, 1973.

Mondt, J. C., SH waves: theory and observations for epicentral distances greater than 90 degrees, *Phys. Earth Planet. Int., 15,* 46-59, 1977.

Mula, A. H., Amplitudes of diffracted long-period P and S waves and the velocities and Q structure at the base of the mantle, *J. Geophys. Res., 86,* 4999-5011, 1981.

Mula, A. H., and G. Müller, Ray parameters of diffracted long period P and S waves and the velocities at the base of the mantle, *Pure Appl. Geophys., 118,* 1270-1290, 1980.

Okal, E. A., I. Application of normal mode theory to seismic source and structure problems; II. Seismic investigations of upper mantle lateral heterogeneity, *Ph.D. Thesis,* California Institute of Technology, 249 pp., Pasadena, 1978.

Okal, E. A., and R. J. Geller, Shear-wave velocity at the base of the mantle from profiles of diffracted SH waves, *Bull. Seismol. Soc. Am., 69,* 1039-1053, 1979.

Olson, P., G. Schubert, and C. Anderson, Plume formation in the D" layer and roughness on the core-mantle boundary, *Proc. XIXth Gen. Assemb. Intl. Un. Geod. Geophys.*, Vancouver, B. C., *V.1*, 6, 1987 [abstract].

Schlittenhardt, J., J. Schweitzer, and G. Müller, Evidence against a discontinuity at the top of D", *Geophys. J. Roy. astr. Soc., 81,* 295-306, 1985.

Tandon, A. N., and H. N. Srivastava, Fault plane solutions as related to known geological faults in and near India, *Ann. Geofis., 28,* 13-27, 1975.

Tanimoto, T., The three-dimensional shear wave structure in the mantle by overtone waveform inversion - I. Radial seismogram inversion, *Geophys. J. R. astr. Soc., 89,* 713-740, 1987.

Woodhouse, J. H., and F. A. Dahlen, The effect of a general aspherical perturbation on the free oscillations of the Earth, *Geophys. J. R. astr. Soc., 53,* 335-354, 1978.

Woodhouse, J. H., and A. M. Dziewonski, Mapping the upper mantle: Three-dimensional modeling of Earth structure by inversion of seismic waveforms, *J. Geophys. Res., 89,* 5953-5986, 1984.

Woodhouse, J. H., and A. M. Dziewonski, Models of the upper and lower mantle from waveforms of mantle waves and body waves, *Eos, Trans. Amer. Geophys. Un., 68,* 356-357, 1987 [abstract].

SEISMIC WAVE SCATTERING AND THE EARTH'S STRUCTURE IN THE LOWER MANTLE

R. A. W. Haddon and G. G. R. Buchbinder

Geophysics Division, Geological Survey of Canada
Energy, Mines and Resources
Ottawa, Ontario K1A 0Y3, Canada

Abstract. The hypothesis of seismic wave scattering by three-dimensional heterogeneity in the lowermost mantle is shown to provide a plausible alternative to recently inferred lower mantle velocity models in which P and/or S velocities increase discontinuously with depth by amounts of 1.5 to 3% at various levels in the lowermost few hundred kilometers of the mantle. In contrast to the stability implied by stratified structures, the required three-dimensional heterogeneity in the lowermost mantle implies strong dynamical activity in this region.

Introduction

Since Bullen's (1949) inference of a distinct layer, which he designated D", at the base of the Earth's mantle, this region has attracted enormous research efforts in attempts to characterize its properties and structure. As has been made clear in a recent extensive review by Young and Lay (1987a), in spite of the enormous efforts made, few (if any) uncontroversial inferences exist concerning any systematic global radial structure inside this region. In many cases conclusions drawn by different authors are opposites with one another. Thus, for example, while some authors have inferred negative P and S velocity gradients with depth (e.g. Cleary, 1974; Doornbos and Mondt, 1979; Ruff and Helmberger, 1982), others (e.g. Mula and Muller, 1980; Dziewonski and Anderson, 1981) have concluded that nearly normal gradients are consistent with observational data in this region. Among the more extreme spherically layered models that have been proposed are those involving sharp discontinuous increases with depth in P velocity (Wright and Lyons, 1981; Wright et al, 1985) and in S velocity (Lay and Helmberger, 1983a, b; Young and Lay, 1986; Wysession and Okal, 1987).

Typical of models involving sharp increases in S velocity with depth in the lower mantle is the model SLHO of Lay and Helmberger (1983a) shown in Fig. 1. This model departs from standard Earth models such as those of Jeffreys (1939) and Dziewonski and Anderson (1981) in having a spherical layer of low (near zero) S velocity gradient between about 2300 and 2600 km depth and a discontinuous increase in S velocity of 2.75% at depth 278 km above the core mantle boundary. Lay and Helmberger have interpreted such models as implying either a phase or compositional change from one side of the discontinuity to the other with the region below this discontinuity (which they identify as D") as a region of (relatively) stable stratification. Young and Lay (1987a) have further recently suggested that this region may be a "refuse pile" for "heavy" mantle heterogeneities. We remark that it appears to have been largely overlooked that the main departures of models such as SLHO from "standard" Earth models are those between about 2300 and 2600 km depth. In this region the standard velocities have been substantially reduced so as to obtain the 2.75% discontinuity without at the same time obtaining velocities that are seriously inconsistent with observational data on S waves diffracted around the core. Thus it is not so much that the region D" in models such as SLHO is anomalous: it is rather the region of substantially reduced velocity that has been implicitly postulated above it. The emphasis in interpreting models such as SLHO should therefore more appropriately be directed towards explaining the elevated layer of low velocity above the postulated discontinuity rather than the relatively normal region below it.

In respect of P velocity variation, it is notable that as shown in Fig. 1 the levels of inferred discontinuous increases (Wright and Lyons, 1981; Wright et al, 1985; and others) are about 100 km lower and are thus seriously inconsistent with those inferred by Lay and Helmberger (1983a) and others for S. Such

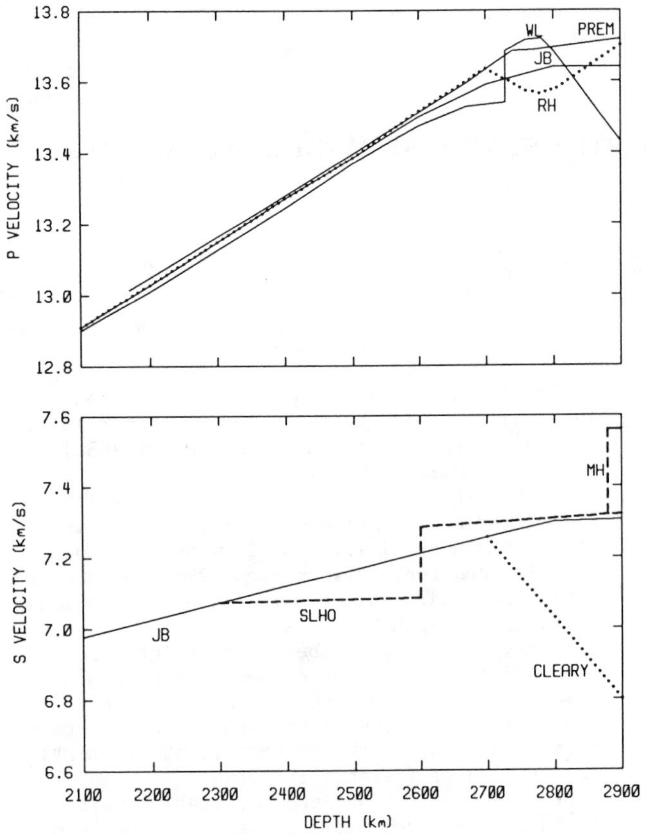

Fig. 1. Comparison of a selection of proposed P and S velocity models for region D". The P velocity models shown are: JB - Jeffreys (1939); PREM - Dziewonski and Anderson (1981); RH - Ruff and Helmberge (1982); WL - Wright and Lyons (1981). The S velocity models shown are: JB - Jeffreys (1939); Cleary (1969); MH - Mitchell and Helmberger (1973); SLHO - Lay and Helmberger (1983).

inconsistencies, together with demonstrations of the inconsistency of both kinds of models with observational data for a sampling of several widely separated mantle paths (Schlittenhardt et al, 1985), clearly establish that if discontinuous increases in P and S velocity with depth do occur, then they are at most local and not global features of Earth structure.

It should be stated that existing controversy over whether velocities in D" are higher or lower than normal clearly throws considerable doubt on either of these possibilities and raises the question as to the existence of any distinct systematic global stratification in this region. This, of course, is not to exclude the possibility of a thin thermal boundary layer at the base of the mantle (Loper and Stacey, 1983; Stacey and Loper, 1983; Loper, 1984). It is simply to question the value of finely detailed spherically symmetric models in a region that is known to have considerable regional variability and 3-D complexity.

While the question of global stratification or otherwise of the lowermost mantle remains unresolved, an increasingly large body of unequivocal evidence continues to amass indicating strong lateral heterogeneity with a variety of scales in this region. On the larger scales the heterogeneity is established by variations in both absolute and relative travel-times of various P and S phases including P, S, PcP, ScS, P(diff), S(diff), PKP, SKS, PKKP and SKKS. At the very long wave lengths we cite as an example the results of tomographic inversions of P travel time data by Dziewonski (1984) indicating large scale lateral variations of velocities in the lowermost mantle, of magnitude 1-1.5% which are 3-4 times greater than those of the mid mantle, in spite of the considerable smoothing involved.

At the other extreme, unequivocal evidence of strong lateral heterogeneity with scales of the order of tens of kilometers is provided by the existence and characteristics of precursors to PKP, P'P' and PKKP (Haddon and Cleary, 1974; Doornbos and Vlaar, 1973; Husebye et al, 1976, Doornbos, 1978). In addition to small scale structures, detailed array data on precursors to PKP also provides clear evidence of systematic laterally heterogeneous structures with scales of the order of some hundreds of kilometers (Haddon, 1982). An example of such evidence is shown in Fig. 2.

With the existence of strong lateral heterogeneity at the base of the mantle very firmly established, it is clearly appropriate to evaluate the effects of such heterogeneity on seismic wave propagation before attributing any particular observed characteristics of P and S signals to postulated radially stratified structure. In what follows it will be shown that the observational data reported by Lay and Helmberger (1983a) and others (Zhang and Lay, 1984; Young and Lay, 1986) can be adequately explained by seismic wave scattering by 3-D heterogeneities. The results to be given clearly demonstrate that while 3-D heterogeneity in the lowermost mantle would undoubtedly involve increases in P and S velocities with depth as well as decreases the former are not absolutely required to satisfy the data. Increases with depth would of course generally act to enhance amplitudes of upwards-scattered waves and such increases most likely play a major role in this regard.

The Observational Data to be Explained

In the following, attention will be focussed on S wave scattering as an alternative to the discontinuously layered models proposed by Lay and Helmberger (1983) and others. Similar P

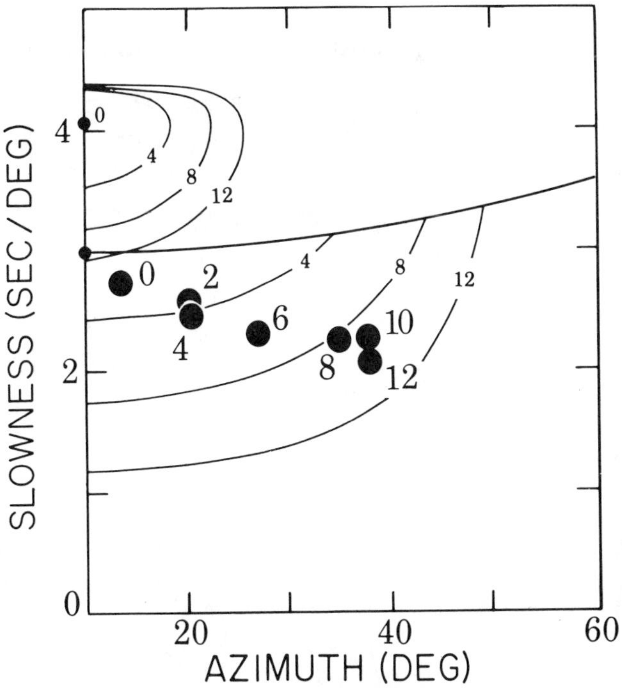

Fig. 2. The black circles show the locations of the centers of dominant power peaks in successive 2 s BEAMAN frames for an example given by Husebye et al. (1976) in their Figure 6. The data refer to a PKIKP precursor wavetrain for an event at a distance of 136° from Norsar. The numbers 0, 2, 4, ... represent time in seconds from the earliest theoretical arrival time for waves singly scattered at the core mantle boundary. Note the systematic migration of observed arrival direction with time. Such behaviour is strongly indicative of scattering from systematic variations in velocity structure of some hundreds of kilometers in extent.

wave scattering may be shown to provide a similar alternative to the discontinuously layered models of Wright and Lyons (1981) and others.

The basic observational data used by Lay and Helmberger (1983a) (to be referred to hereafter as LH),and their co-workers, consist of long period N-S and E-W seismometer recordings from which "rotated" seismograms corresponding to the transverse horizontal component of particle motion were constructed. For an ideal spherically symmetrical Earth, such rotated seismograms would depend only on SH motion and would not contain any contributions from P-waves nor from vertically polarized phases such as SKS. The particular evidence for the velocity discontinuity inferred by LH consists of certain small arrivals which they have designated as the Scd phase. These arrivals generally manifest themselves as small irregular variations in the shapes of SH waveforms for epicentral distances between about 70 and 95°. In general, as discussed by Haddon and Buchbinder (1986), scrutiny of Lay and Helmberger's entire set of published data shows that the Scd arrivals are highly irregular functions of both source and receiver location, though a few clear examples of coherent arrivals do exist for a few small groups of relatively closely spaced receivers. One example of the latter is shown in Fig. 3.

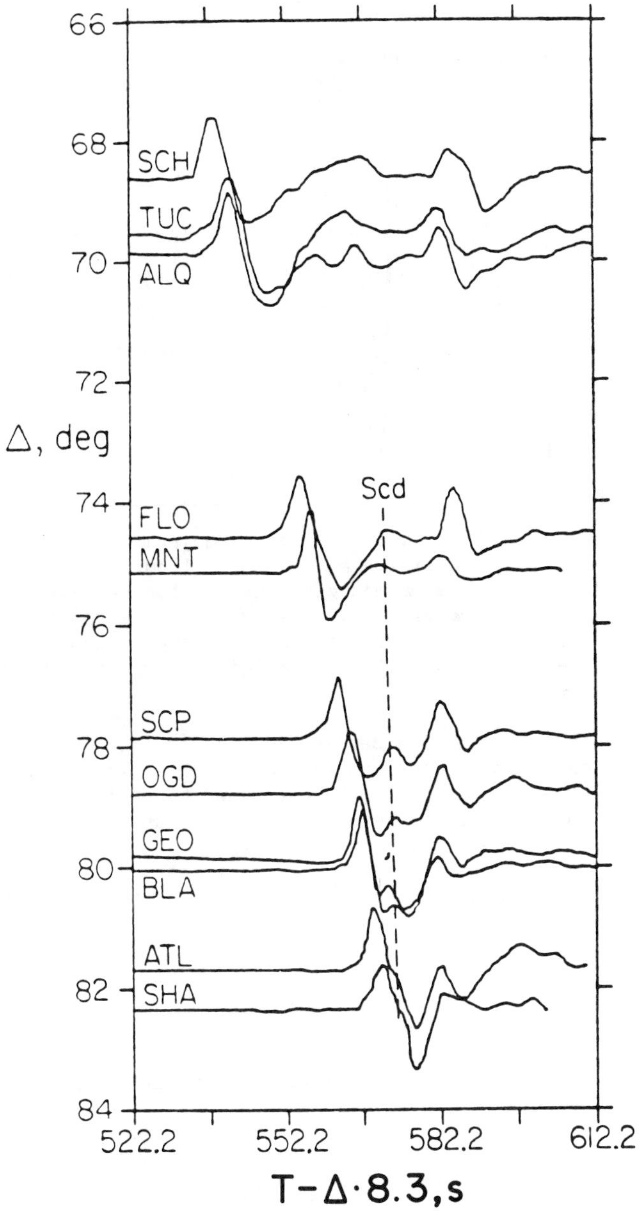

Fig. 3. Figure 6 from Lay and Helmberger (1983a). showing an example of a small relatively coherent arrival between S and ScS which they have designated as the Sed phase.

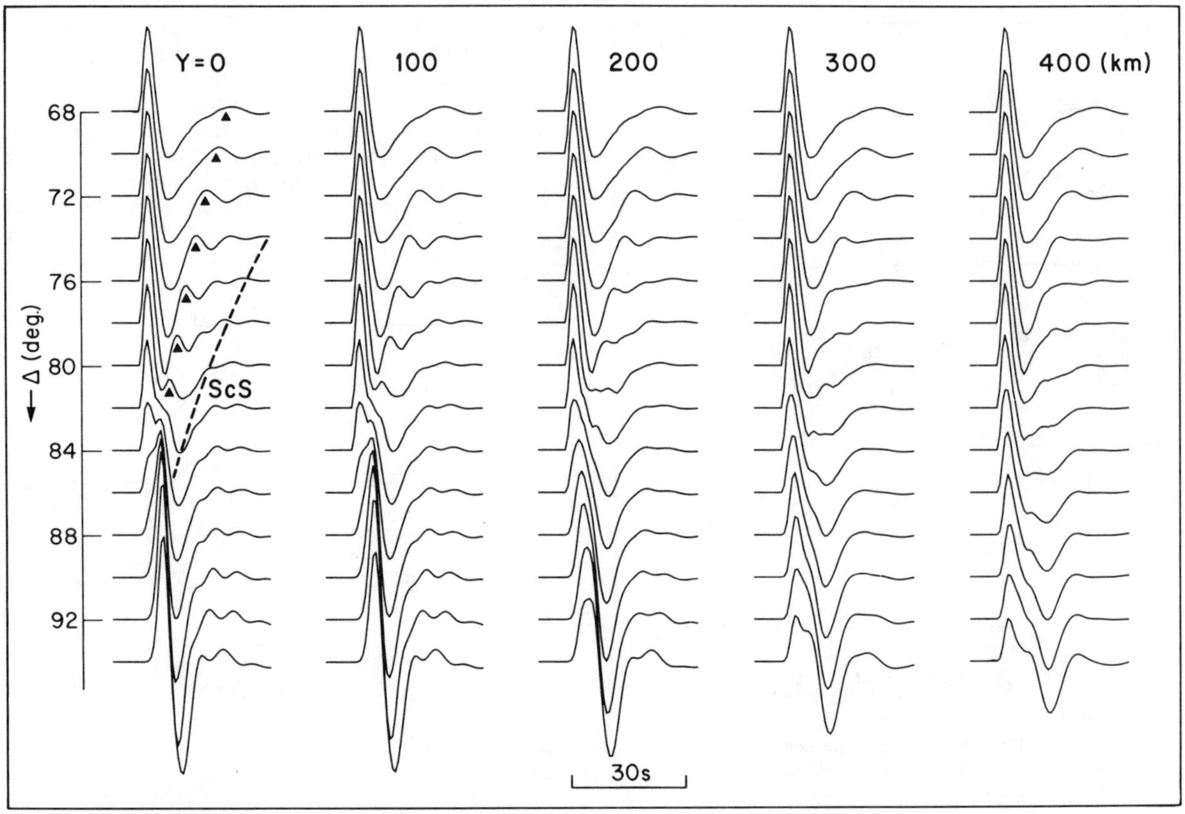

Fig. 4. Synthetic seismograms adapted to a spherically symmetrical Earth model containing a vertical cylindrical velocity anomaly as described in the text. The five panels give results for receivers having lateral offsets of 0, . . . 400 km from the diametral plane containing the source and the axis of the anomaly. Scattered wave arrivals corresponding to Scd observations (triangles) and arrival times for ScS (dashed line) are indicated in the first panel.

In the following section it will be shown that examples of coherent arrivals such as those shown in Fig. 3 can be explained by postulating very simple 3-D heterogeneous structures. Examples of less coherent arrivals can obviously be explained by the more complicated 3-D heterogeneous structures which would be expected to occur in the real Earth.

Scattering Effects of Lateral Heterogeneities in D"

Fig. 4 shows synthetic seismograms which were computed for a single semi-infinite axi-symmetric cylindrical velocity anomaly embedded in a homogeneous isotropic medium with S-wave velocity 7.2 km/sec (shown schematically in Fig. 5).

The radius of the cylindrical anomaly was taken to be 350 km and the velocity difference was taken to vary quadratically with distance from a maximum of 0.5 km/sec on its axis to zero on its surface. A spherically symmetrical source located at a point in the plane of the flat end of the anomaly at a distance of 4000 km from its axis was assumed. The synthetic seismograms shown are for a distance of 8000 km from the source. They were computed using the parabolic wave equation method of Claerbout and Johnston (1971) generalized to apply for 3-dimensional wave propagation. In Fig. 4 appropriate allowances have been made for the curvature of rays caused by slow increase in velocity with depth so that the results and the distance scale in Fig. 4 refer approximately to a standard spherically symmetrical Earth model with an axially symmetric velocity anomaly having a vertical axis and extending upwards to a level about 400 km above the core-mantle boundary (cmb). We should mention that we have not made any attempt to model the reflection ScS from the cmb. This phase, however, does not have any effect on the waveforms shown prior to the appropriate ScS arrival time (the latter is indicated approximately by the dashed line in

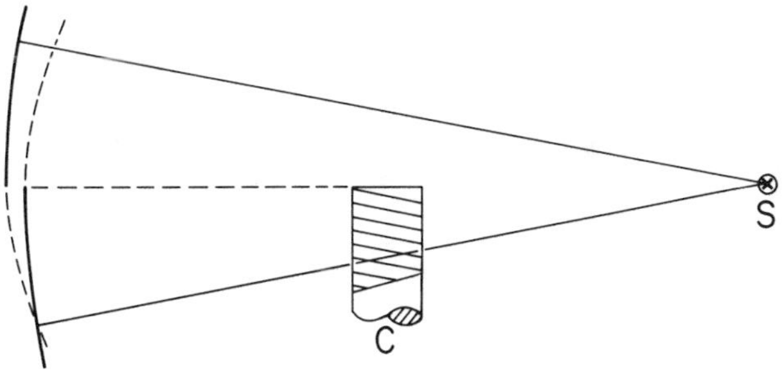

Fig. 5. Schematic diagram showing source S and semi-infinite cylindrical velocity anomaly C in an otherwise homogeneous medium. Wavefronts corresponding to rays from the source are indicated by full lines. The dashed line continuations indicate diffraction/scattering effects that produce the arrivals corresponding to Scd observations in Fig. 4.

the first panel of the figure). The perturbation of the wavefield caused by the anomaly varies both with epicentral distance and azimuth. The seismograms in the first panel in Fig. 4 correspond to cases in which both the source and receivers, as well as the axis of the anomaly, all lie in the same diametral plane of the model. The results in the second to fifth panels refer to receivers laterally offset from this plane by 100, 200, 300 and 400 km, respectively.

The results in Fig. 4 show distinct coherent secondary arrivals with arrival times consistent with observed arrival times of the Scd phase (cf. Fig. 3). Note that the secondary arrivals in the first two panels of Fig. 4 are relatively sharp for epicentral distances between about 82 and 76°. With decreasing distance, however, the arrivals in these and all of the other panels increasingly spread out with decreasingly lower frequency content. This is an important characteristic of all of the Scd data published by LH which is not accounted for by models such as SLHO. Note also that for distances near 83° and beyond the waveforms exhibit the so called "crossover" characteristics and pulse broadening observed and discussed by LH.

Important is the fact that with our model, the waveforms vary as functions of source and receiver location with respect to the anomaly. This is a crucial characteristic of the data. As particular examples of this we point to the waveforms observed at stations VAL and TRI in Figs. 25 and 26 of LH which show clear Scd signals. The corresponding waveforms at the station STU, which is situated between VAL and TRI at almost the same epicentral distance, shows no sign of Scd whatsoever. Further examples abound in the other figures given by LH. Extreme variability of Scd waveforms observed at different instruments only 50 km apart in the Graefenberg array have also been exhibited by Cormier (1985). These important characteristics are consistent with theoretical results for models that are laterally heterogeneous both in the lower and upper mantle, but inconsistent with those for spherically symmetrical models such as SLHO.

The laterally inhomogeneous models we propose can also accommodate the wide ranges of observed slopes of diffracted ScS. Observed slopes around 8.4 sec/deg are accommodated by ray paths which sample regions of the lower mantle in which the average velocity is close to the Jeffreys values. Observed slopes near 9 sec/deg and greater and associated travel time delays (of up to 12 sec at Δ = 105°) (Cleary, 1969; Bolt et al. (1970), Bolt and Niazi (1984)), evidently require substantial reduction from the Jeffreys velocity values somewhere along the associated paths taken by the waves. The calculated travel times for laterally inhomogeneous models are evidently strongly dependent on the locations of the velocity anomalies with respect to the source and receiver. It is therefore to be expected that the above extreme deviations in slope would occur only relatively infrequently, which is what is observed.

Fig. 6 shows, for example, rays for a model in which the velocity is decreased linearly with depth from the Jeffreys values in a local anomalous region by from zero at level Z = 2300 km to 0.5 km/sec at the cmb and the corresponding travel times obtained. As indicated in the figure such an anomaly produces a travel time delay of 12 s. at 100° when the source and receiver are located with respect to the anomaly as shown. The associated slope of the T - Δ curve for diffracted ScS for Δ > 90° is 9.5 sec/deg. Because the majority of observed slopes are near 8.4 sec/deg such extreme anomalies need occur only in a very small proportion of the total cmb region. For example, the anomaly shown in Fig. 5 covers less than 3% of the surface area of the core.

Conclusions

We have demonstrated using an extremely simple model that arrivals having all of the principal observed characteristics of the Scd data published by Lay and Helmberger (1983a) and their co-workers can be produced by wave scattering due to 3-D heterogeneities in the lowermost mantle. Similar arrivals are also produced by scattering of P waves. We should make it clear that we do not propose that 3-D heterogeneities in the lowermost mantle are likely to have any such simple geometrical structure as the one we have assumed. Nor do we propose that the heterogeneities are likely to be such that velocities would everywhere decrease with depth. In this connection we

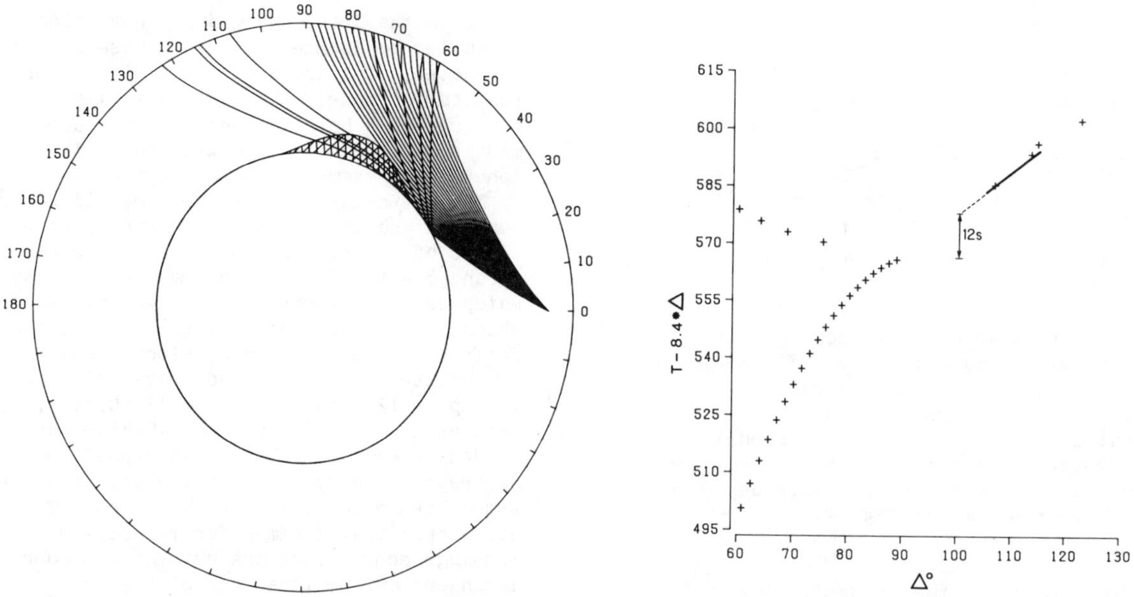

Fig. 6. Effect of a low velocity anomaly on the travel times and slope dT/d Δ of diffracted SH. The model shown is spherically symmetrical except in the locally anomalous region indicated where the S velocity is reduced linearly with depth from the upper boundary of the anomaly to the cmb. The radial velocity gradient inside the anomaly is constant and corresponds to a reduction in the Jeffreys velocities by from zero at Z = 2300 km to 0.5 km/sec at the cmb. For sources and receivers in the locations shown the anomaly causes a travel time delay relative to the J.B. tables of 12 sec. at 100° and an increase in dT/d Δ from 8.4 sec/deg to 9.5 sec/deg for Δ > 100°.

remark that we have examined scattering effects of other kinds of simple velocity anomalies including, for example, low velocity ellipsoidal inclusions. Many such velocity anomalies, including those likely to be associated with the formation of plumes (Olsen et al., 1987), cause waveform variations and secondary arrivals having characteristics consistent with the Scd observations. Of particular interest, however, is the fact that the model we have chosen to demonstrate the viability of the proposed scattering mechanism, nowhere has sharply increasing velocity with increasing depth.

Our results, taken in conjunction with the results of Schlittenhardt et al. (1985) and uncertainties such as those pointed out by Cormier (1985), show that Lay and Helmberger's (1983a,b) and Young and Lay's (1987b) inferred discontinuity or zone at the top of region D" in which the S velocity increases sharply by about 3% from above to below is not demanded by their data. Seismic wave scattering by three dimensional heterogeneity at the base of the mantle has been shown to provide a plausible alternative, consistent with almost all other independent lines of evidence.

Acknowledgments. The authors wish to thank their colleagues P.W. Basham and M.J. Berry for their helpful comments on the manuscript and T. Lay for permission to reproduce Fig. 3. Geological Survey of Canada Contribution No. 30987.

References

Bolt, B.A., Niazi, M., S velocities in D" from diffracted SH-waves at the core boundary. Geophys. J. R. Astron. Soc. 79: 825-834, 1984.

Bolt, B.A., Niazi, M., Somerville, M.R., Diffracted ScS and the shear velocity at the core boundary. Geophys. J. R. Astron. Soc. 19: 299-305, 1970.

Bullen. K.E., Compressibility-pressure hypothesis and the Earth's interior, Mon. Not. R. Astron. Soc. 5: 355-368, 1949.

Claerbout, J.F. and Johnson, A.G., Extrapolation of time dependent waveforms along their path of propagation, Geophys. J.R. astr. Soc., 26: 285-293, 1971.

Cleary, J.R., The S velocity at the core-mantle boundary, from observations of diffracted S, Bull. Seismol. Soc. Am. 59: 1399-1405, 1969.

Cleary, J.R., The D" region. Phys. Earth Planet Inter. 19: 13-27, 1974. Cormier, V.F., Some problems with SKS and ScS observations and implications for the structure at the base of the mantle and the outer core, J. Geophys., 57: 14-22, 1985.

Doornbos, D.J., On seismic-wave scattering by a rough core-mantle boundary, Geophys. J.R. Astron. Soc. 53: 643-662, 1978.

Doornbos, D.J., Vlaar, N.J., Regions of seismic wave scattering in the Earth's mantle and precursors to PKP. Nature Phys. Sci. 243: 58-61, 1973.

Doornbos, D.J., Mondt, J.C., P and S waves diffracted around the core and the velocity structure at the base of the mantle, Geophys. J. R. Astron. Soc. 57: 381-395, 1979.

Dziewonski, A.M., Mapping the lower mantle: determination of lateral heterogeneity in P velocity up to degree and order 6. J. Geophys. Res. 89: 5929-5952, 1984.

Dziewonski, A.M., Anderson, D.L., Preliminary reference Earth model, Phys. Earth Plant. Inter. 25: 297-356, 1981.

Haddon, R.A.W., Evidence for inhomogeneities near the core-mantle boundary, Phil. Trans. R. Soc. Lond. A306: 61-70, 1982.

Haddon, R.A.W. and Cleary, J. R., Evidence for scattering of seismic PKP waves near the mantle-core boundary, Phys. Earth Planet. Inter. 8: 211-234, 1974.

Husebye, E.S., King, D.W., and Haddon, R. A. W., Precursors to PKIKP and seismic wave scattering near the mantle-core boundary, J. Geophys. Res. 81: 1870-1882, 1976.

Jeffreys, H., The times of P, S, and SKS and the velocities of P and S. Mon. Not. R. Astron. Soc. 4: 498-533, 1939.

Lay, T., Helmberger, D.V., A lower mantle S-wave triplication and the shear velocity structure of D". Geophys. J. R. Astron. Soc. 75: 799-838, 1983a.

Lay, T., Helmberger, D.V., The shear-wave velocity gradient at the base of the mantle. J. Geophys. Res. 88: 8160-8170, 1983b.

Loper, D.E., The dynamical structures of D" and deep plumes in a non-Newtonian mantle, Phys. Earth Planet. Inter. 39: 57-67, 1984.

Loper, D.E., Stacey, F.D. The dynamical and thermal structure of deep mantle plumes. Phys. Earth Planet. Inter. 34: 304-317, 1983.

Mula, A.H., Muller, G., Ray parameters of diffracted long period P and S waves and the velocities at the base of the mantle. Pure Appl. Geophys. 118: 1272-1292, 1980.

Olson, P., Schubert, G. and Anderson, C, Plume formation in the D" layer and the roughness of the core-mantle boundary, Nature, 327: 409-413, 1987.

Ruff, L. J., Helmberger, D.V., The structure of the lowermost mantle determined by short-period P-wave amplitudes, Geophys. J. R. Astron. Soc. 68: 95-119, 1982.

Schlittenhardt, J. , Schweitzer, J., Muller, G., Evidence against a discontinuity at the top of D". Geophys. J. R. Astron. Soc. 81: 295-306, 1985.

Stacey, F.D., Loper, D.E., The thermal boundary-layer interpretations of D" and its role as a plume source, Phys. Earth Planet, Inter. 33: 45-55, 1983.

Wright, C., Lyons, J.A., Further evidence for radial velocity anomalies in the lower mantle, Pure Appl. Geophys. 119: 137-162,1981.

Wright, C., Muirhead, K. J., Dixon, A.E., The P wave velocity structure near the base of the mantle. J. Geophys. Res. 90: 623-634, 1985.

Wysession, M.E., Okal, E. A. , Modal synthetics for diffracted SH: evidence for a high velocity zone at the base of the mantle, IUGG Abstracts VI, 1987.

Young, C.J., Lay, T., Evidence for a shear velocity discontinuity in the lowermost mantle beneath Indian and the Indian Ocean, Eos, Trans. Am. Geophys. Union. 67: 311 (Abstr.), 1986.

Young, C.J., Lay, T., Comments on wave propagation effects and the Earth's structure in the lower mantle, Geophys. Res. Letters, 14: 562-565, 1987a.

Young, C.J. and Lay, T., The core-mantle boundary, Ann. Rev. Earth Planet. Sci., 15,: 25-46, 1987b.

Zhang, J., Lay, T., Investigation of a lower mantle shear wave triplication using a broadband array, Geophys. Res. Lett. 11: 620-623, 1984.

PLANETARY SCALE FLOW IN THE EARTH'S CORE AND GEODETIC OBSERVATIONS

J. B. Merriam

University of Saskatchewan, Saskatoon, Saskatchewan S7N-0W0

Abstract. The westward drift of the geomagnetic field implies fluid velocities in the outer core of $4 \times 10^{-4}\, m\, s^{-1}$. The pressure associated with this flow is small (about 10^4 Pa) but it is shown that this is sufficient to produce observable deformations of the mantle. Changes in J_2, polar motion, and the length-of-day, are all measured with sufficient accuracy that these data could contain a component due to pressure anomalies, of this magnitude, at the core-mantle boundary. Relations between pressure at the core-mantle boundary and geodetic quantities such as J_2, polar motion, and the length-of-day are derived and the unexplained residual in each is used to set an upper limit on the decade-scale pressure fluctuations at the core-mantle boundary. The polar motion observations supply the most severe upper bound ($< 2 \times 10^3$ Pa) but satellite observations of J_2 could soon prove to be more useful. It is evident that not more than 10% of the decade changes in the length-of-day can be due to pressure core-mantle coupling.

Introduction

Fluid motions in the Earth's core are thought to produce the geomagnetic field by dynamo action, but the solution to the dynamo problem is hampered by our lack of knowledge of the motions in the core. The scant information we do have is the result of observations of the secular variation of the magnetic field at the surface, and the assumption that some of these changes monitor fluid motion at the core-mantle boundary (cmb) by virtue of the magnetic field being "frozen" into the fluid by the high conductivity of the core. The westward drift of the non-dipole field, by about $0^o.2\, yr^{-1}$, has long been interpreted to imply a fluid velocity near the core-mantle boundary of $4 \times 10^{-4}\, m s^{-1}$. Recently, several attempts have been made to derive more detailed knowledge of the core motions from the secular variation [Voorhies, 1986, Whaler, 1986, Le Mouel et al., 1985].

The purpose of this paper is to show that some geodetic measurements may supply further information on the large scale motions in the core. This information could be of great use in supporting fluid flow models derived from the secular variation. Of course, it is impossible to associate an observed variation in any geodetic measurement with a change in the pressure regime at the cmb unless perhaps some correlation is found with the secular variation. The purpose of this paper is not to search for such correlations, but merely to define the relationship between pressures at the cmb and geodetic measurements, and thereby to indicate what magnitude of pressure is necessary to force a detectable change in some quantity measured at the Earth's surface.

Fluid motions in the Earth's core must of course be accompanied by lateral pressure gradients that express a departure from hydrostatic equilibrium. Because the energy available to drive fluid motions is small the flow is not vigourous, and large pressure anomalies are not anticipated. However, it is shown here that surprisingly small pressure anomalies at the core-mantle boundary can produce detectable deformations of the mantle. There are numerous lines of evidence that suggest that these pressure anomalies could be about $10^4\, Pa$. The accuracy of observations of: the length-of-day (lod), J_2 , and polar motion are such that these would show a response from pressure anomalies of that magnitude. Other possible observations: tilts, sea-level, strains, and gravity are probably not sensitive enough.

There are several arguments that can be used to show the range of pressures that can be expected. Wahr[1987] has used satellite data on the second degree tesseral harmonics of the Earth's gravity field to set an upper bound on the tilt of the core's rotation axis with respect to the mantle's symmetry axis. This in turn sets an upper bound on the equatorial torques the core exerts on the mantle, and hence an upper limit on the long wavelength pressure anomalies of $1.5 \times 10^4\, Pa$.

The magnitude of the pressure gradients that can exist along the cmb can also be estimated by assuming that the fluid flow in the core is in geostrophic balance, with the Coriolis force balancing the pressure gradient.

$$2\vec{\Omega} \times \vec{U}\rho = -\vec{\nabla}P$$

Where $\vec{\Omega}$ is the rotation rate, $\vec{V}$ is the fluid flow velocity , ρ is the density and P the flow pressure. Using the westward drift as an estimate of the velocity, and assuming the pressure field has a wavelength of the circumference of the core, then anomalies in pressure of $\sim 10^4\, Pa$ might be expected.

It is difficult to detect decade scale anomalies in this way because the conductivity of the mantle masks magnetic fluctuations on time-scales shorter than a couple of decades. The decade variations in the length-of-day (lod) however, imply that decade motions, of some scale must exist, along with decade pressure variations. Lacking evidence to the contrary, the figure of $10^4\, Pa$, which derived from secular time-scales, may be viewed as at least a plausible range for decade-scale variations in pressure along the cmb.

Deformation of the Mantle by Pressure at the Core-Mantle Boundary

Deformations of the Earth by body force potentials, or surface loads, are conveniently handled by Love numbers. For the problem considered here, we need quivalent numbers which characterize the

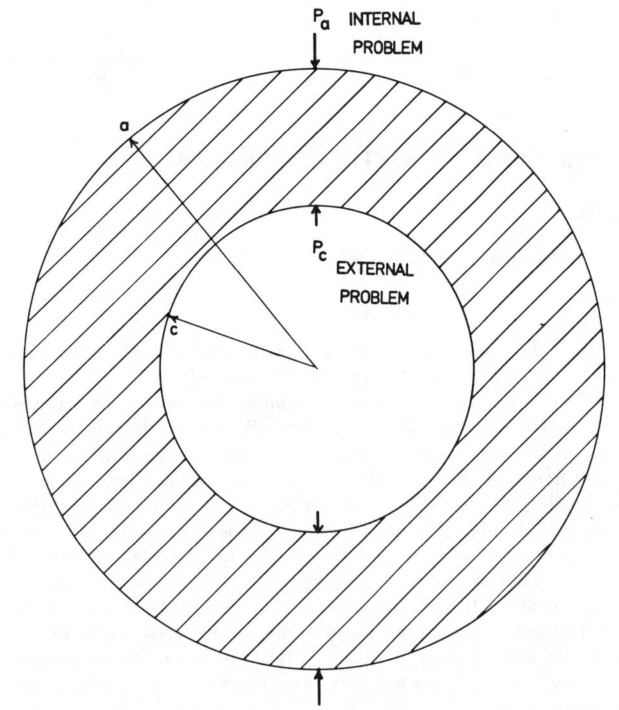

Fig. 1. The thick shell problem has two classes: the internal problem, in which the shell is internal to the loaded surface, and the external problem, in which the shell is external to the loaded surface. The former corresponds to the usual tidal load problem with a load P_a on the surface of radius a, and the latter to the problem under consideration here, with a load P_c on the surface of radius c.

response of the mantle to a harmonic load on the cmb. Approximate solutions to the problem can be obtained by examining the general problem of the thick spherical shell, for which well-known analytic solutions exist.

The general solution to the thick shell problem has four linearly independent solutions, and can be divided into two classes. The first class is called the internal solution, where the shell is internal to the loaded surface, and the second class is called the external solution, where the shell is external to the loaded surface (Figure 1). The former is the usual tidal load problem and the latter is dealt with here. Although self-gravitation and radial stratification are important, they can be simply modeled by calibrating the solutions to the thick shell internal problem so that they agree with observed Love numbers. The solutions to the external problem should then be within a few percent of the true solution.

The differential equations describing the thick shell problem are a fourth order set which can be shown to be a subset of the equations describing the response of a gravitating, radially stratified, elastic sphere. The formulation in Merriam [1988] can be improved by requiring all four solutions together to satisfy the boundary conditions, in which case the radial displacements in the internal and external problems (u_i and u_e respectively) are found to be:

$$u_i = \left[0.378\left(\frac{r}{a}\right)^3 - 1.482\left(\frac{r}{a}\right) - 0.344\left(\frac{r}{a}\right)^{-2} + 0.045\left(\frac{r}{a}\right)^{-4}\right]\frac{aP_a}{2\mu}$$

$$u_e = \left[-0.101\left(\frac{r}{a}\right)^3 + 0.347\left(\frac{r}{a}\right) + 0.191\left(\frac{r}{a}\right)^{-2} - 0.019\left(\frac{r}{a}\right)^{-4}\right]\frac{aP_c}{2\mu}$$

where P_a, is a second degree harmonic of pressure at the surface and, P_c a second degree harmonic of pressure at the cmb. By adjusting the value for the rigidity μ, so that the internal solution gives the observed radial displacement body force Love number, and has a potential characterized by the observed potential Love number, we can effectively calibrate this simple model to respond in the same way as a gravitating radially heterogeneous model. The calibrated value rigidity μ, is 4×10^{11} Pa.

From the displacement solutions we can easily derive approximate results for the change in potential at the surface and the result for the internal problem can be compared with the observed Love number. The change in potential immediately gives the change in J_2, and the rotation perturbations m_1, m_2, and m_3.

$$\begin{aligned} P_c &= -0.78 \times 10^{13}\,\Delta J_2 \\ P_c &= 3.9 \times 10^{12}\, m_3 \\ P_c &= 1.9 \times 10^{10}\, m_1, m_2 \end{aligned} \tag{1}$$

These are the relations between the second degree harmonics of flow pressure, in Pa, at the cmb and the change in J_2 and m_1, m_2, m_3. The sensitivity in all three of these is such that they should show a response to pressure changes of much less than 10^4 Pa at the cmb.

Changes in the Length-of-Day

Changes in the length-of-day (lod) of a few ms, that persist for from several years to several decades, are referred to as the decade variations in the lod. These fluctuations have usually been attributed to the exchange of angular momentum with the fluid core because of the scale of the phenomenon. The presence of lateral pressure variations, of harmonic degree two, at the cmb, raises the possibility of accounting for at least some of the decade variations in lod by the action of these pressure anomalies on the inertia tensor of the Earth.

Figure 2 shows the decade variations in lod as compiled by Morrison [1979]. A secular rate of $21.1 \pm 5.6 \times 10^{-3}$ $ms\,yr^{-1}$, due to tidal friction ($27.2 \pm 5.6 \times 10^{-3}$ $ms\,yr^{-1}$), post glacial rebound ($-6.1 \pm 0.5 \times 10^{-3} ms\,yr^{-1}$), and atmospheric tides ($-.3 \times 10^{-3} ms\,yr$), has been removed. If we were to attribute all of the remainder to changes in pressure at the cmb, then the relation between such

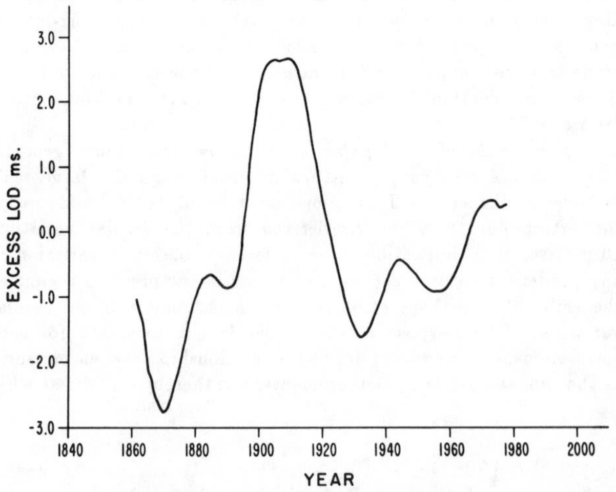

Fig. 2. The excess length-of-day (LOD) in ms versus the year from 1861 to 1978, from Morrison (1979). A linear rate of change has been removed.

pressures and the lod (1), requires changes in pressure of about $3 \times 10^5 Pa$. This is an order of magnitude larger than the pressures suggested by a geostrophic westward drift. However, the precision with which the lod can be measured, suggests that pressure variations of only 10^3 Pa should imprint a signal on the lod.

Not all of the decade variations need to be explained in this way because there are contributions from circulation in the oceans and atmosphere. Annual variations in atmospheric angular momentum of 10^{26} $kg\,m^2\,s^{-1}$ are observed superimposed on a D.C. angular momentum of about the same magnitude. It is premature to judge whether or not there are decade-scale changes in atmospheric angular momentum of similar size but the linear trend between 1976 and 1985 was sufficient to increase the lod by 0.03 $ms\,yr^{-1}$. Over the same period the observed drift in lod was 0.17 $ms\,yr^{-1}$. Over this restricted period the atmosphere thus appears to affect the decade variations in lod at about the 20% level. This is in agreement with the estimate of Lambeck and Cazenave [1974].

In principle, the axial angular momentum of the oceans should be easier to monitor than that of the atmosphere because there is only one narrow band where an East-West current can circumnavigate the globe and this is between South America and Antarctica. However, the Antarctic circumpolar current is not well studied. Whitworth and Petersen [1985] estimate the mean transport of the current through the Drake passage to be 10^8 $m^3\,s^{-1}$, with fluctuations of comparable magnitude. The relative angular momentum is then 10^{25} $kg\,m^2\,s^{-1}$, or about 10% of the fluctuations in angular momentum represented by decade changes in lod. With atmospheric and oceanic effects accounted for, the decade variations in lod are then compatible with pressure fluctuations as small as 10^5 Pa, or as large as 10^6 Pa . These are one and two orders of magnitude larger than the pressure anomalies anticipated from the westward drift. It seems unlikely that one could invert rotation data for core-mantle pressure anomalies and the implication is that if flow pressures at the cmb amount to only 10^4 Pa, then not more than 10% of the decade variations in lod could be produced by pressure core-mantle coupling.

Decade Polar Motion

Polar motion is dominated by the 14 month Chandler wobble, a roughly circular, prograde motion, of the instantaneous rotation axis through the body of the solid Earth. The diameter of the Chandler pole path is about 0.2″, or 10^{-6} rad, with fluctuations of similar size. Accompanying the Chandler wobble is a forced annual wobble caused by seasonal shifts in atmospheric and hydrologic mass. Its amplitude is generally about half the size of the Chandler component. In addition, there is a broad spectrum of shorter period and longer period polar motion. For example, it has long been recognized that the mean rotation pole, or axis of principal moment of inertia, is migrating through the geographic coordinate system, and this motion is commonly referred to as true polar wander, or secular polar motion. During this century the secular pole position has migrated from the Conventional International Origin (the mean position of the pole in 1905) roughly in the direction 76° west of Greenwich [Dickman,1981], at a rate of about 1.65×10^{-8} $rad\,yr^{-1}$, or a distance of somewhat more than a Chandler pole path diameter in 87 years. Accompanying the linear drift is an irregular, decade-scale motion with an amplitude of about $10^{-7} rad$.

The linear part of the drift can be explained as the residual viscoelastic rebound of the mantle following the last deglaciation. Nakiboglu and Lambeck [1980], and Peltier [1983], find that the direction and rate of secular polar motion this century, are compatible with the distribution of the ice load, and the viscosity of the mantle, inferred from glacio-isostatically raised beaches. The residual, after removal of the linear drift, appears to be a broad band spectrum with a line peak at about thirty years [Wilson and Vicente, 1980]. This motion is known as the Markowitz wobble [Markowitz, 1960] and has been described by Dickman [1981] as a highly elliptical retrograde pole path with amplitude about 10^{-7} rad in either component. It is apparent as a motion nearly transverse to the linear drift. Its existence has been doubted because its small amplitude is near the estimated accuracy of even the revised International Latitude Service data of Yumi and Yokoyama[1981],and it is not apparent in the data of the Bureau International de l'Heure.

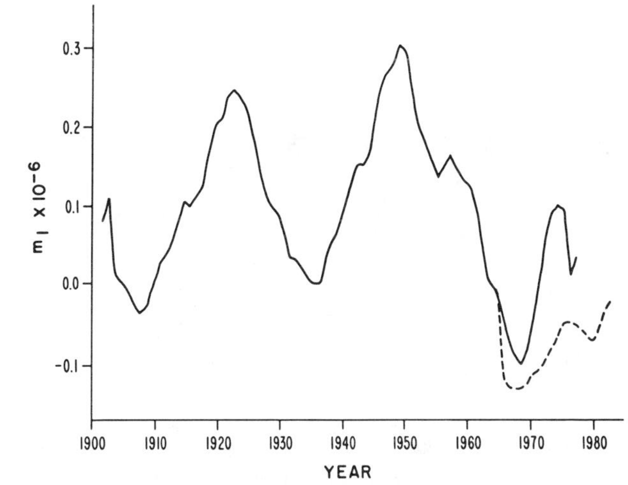

Fig. 3. Secular polar motion in m_1. The Chandler and annual terms have been removed, along with a linear drift. The solid line is the homogeneous ILS series and the dashed line is the BIH series.

Figure 3 shows the secular polar motion of m_1 in the ILS data. A linear drift has been removed. If we were to attribute this motion to changes in pressure at the cmb, then the variations in pressure are seen to be no larger than $2 \times 10^3 Pa$. Assuming the BIH data (the dashed lines at the right hand side) does not have a large Markowitz component, then pressure anomalies of 10^3 Pa are indicated. These are smaller than the estimate based on the westward drift. It may then be argued that, either a substantial fraction of decade polar motion is due to changes in the pressure regime at the cmb, or pressure variations in the core at decade period are smaller than the westward drift indicates. The precision of modern polar motion services (10^{-8} rad) suggests that pressure variations as small as 100 Pa could be detected.

Wahr[1987] uses estimates of the Earth's C_{21}, S_{21} potential coefficients to argue that the rotation axes of core and mantle may be misaligned. The upper bound on the misalignment then supplies an upper bound of 10^3 Pa for the pressure variations on the cmb, if the misalignment is due to a torque alone, and not to the deformation of the mantle as well. This result is consistent with that found here. Wahr anticipates that with future improvements to the satellite data the upper bound on observable pressure anomalies might be only 4×10^2 Pa.

Changes in J_2

The coefficient J_2, in a spherical harmonic expansion of the Earth's gravitational potential, is a measure of the difference between the polar and equatorial moments of inertia, C and A respectively. Satellite orbits will precess at a rate proportional to J_2. This has

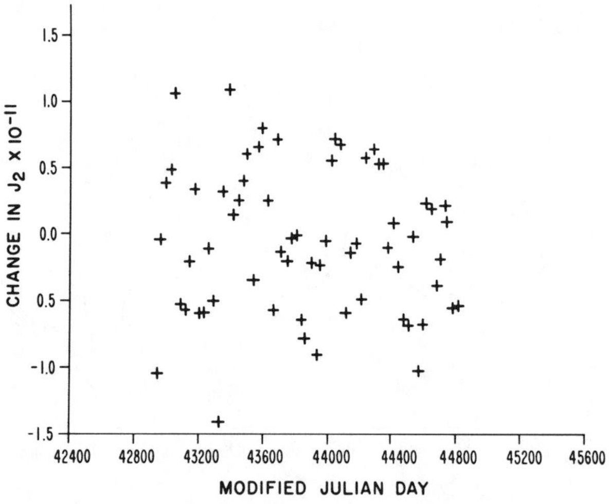

Fig. 4. The residuals in the ΔJ_2 data of Rubincam, after tidal and linear terms have been removed.

been used for thirty years to measure J_2 with the result that it is now known much more accurately than is the total mass of the Earth. A change in J_2 will produce an accelerated precession rate proportional to the rate of change of J_2. Rubincam [1984], and Yoder et al. [1983] have exploited this relationship to measure a secular decrease in J_2 of about $-10^{-18}\ yr^{-1}$. Yoder at al. [1983], Peltier[1983] and Rubincam [1984] have attributed this rate to the on-going viscoelastic adjustment of the mantle to the removal of the Pleistocene ice sheets. Inversions of this rate for mantle viscosity are roughly corroborative of the viscosity inferred from raised and submerged beaches, and from secular polar motion.

Figure 4 shows Rubincam's data for the changes in J_2, after removal of mean, linear, and tidal terms. These residuals have power mostly at periods less than two years. The absence of longer period variations may be real but it could also be due to the modeling process, which may absorb longer period variations into some of the other parameters modeled in the orbit analysis.

The ΔJ_2 values, taken at face value as an indication of the sensitivity of these measurements, suggest pressure fluctuations of less than 100 Pa. However, as noted above, there may in fact be decade variations in J_2, removed by the modeling process, that would suggest much larger pressures.

The problem with this data set is that it is so short that it is impossible to say if the $10^{-18}\,rad\,s^{-1}$ decrease in J_2 is really secular, and therefore more likely a measure of post-glacial rebound, or part of a decade style variation, which might be attributed to fluctuations in cmb pressure. If it is the latter than the pressure at the cmb would be changing by at most 500 $Pa\,yr^{-1}$. On a decade scale this would imply an upper bound on pressures similar to that inferred from decade polar motion, ie less than 10^4 Pa. It seems likely that most of the $-10^{-18}\ rad\,s^{-1}$ is due to post-glacial rebound simply because that rate is consistent with other measured rates and therefore the residual decade-scale variations in J_2 are likely to require a variation in cmb pressure of much less than $10^4\,Pa$. Many more years of Lageos data will be required to settle the issue but the promise of the data is clear.

A pressure variation, that will cause a change in J_2, will also change the lod. A secular change in pressure of 500 $Pa\,yr^{-1}$ will change the lod by -11 $\mu sec\,yr^{-1}$. During the span of the Lageos data presented here, the observed secular rate in lod (that is over seven years), after removal of tidal and wind effects, was about $+100$ $\mu sec\,yr^{-1}$, ie much larger, and of opposite sign to the change in lod implied by ΔJ_2. The tentative conclusion is that only a small fraction, 10% , of the decade changes in lod may be attributed to fluctuations in pressure at the cmb.

Decade Trends in Sea-Level, Tilt, and Gravity

Using (1) it is possible to derive relations between cmb pressure and tilts, sea-level, strains, and gravity.

$$\begin{aligned}
\text{tilt} &= 2.8\times10^{-13}(rad/Pa)\times P_c &&\approx 4\times10^{-9}\\
\text{sea level} &= 2.1\times10^{-6}(m/Pa)\times P_c &&\approx 2\,cm\\
\text{strain} &= 1.2\times10^{-12}(/Pa)\times P_c &&\approx 1\times10^{-8}\\
\text{gravity} &= 1.0\times10^{-3}(\mu gal/Pa)\times P_c &&\approx 10\,\mu gal
\end{aligned}$$

where the column of approximate numbers, on the right, gives the reponse to a pressure anomaly of $10^4\,Pa$ at the cmb. However, it appears that none of these offer much scope for deriving any information on pressures at the cmb.

From the above equations, one might expect tilts of about $4\times10^{-9}\,rad$ (0.002") from pressure fluctuations of $10^4\ Pa$, but there are no good tilt measurements, of long enough duration and stability, to be useful. One possibility is the z term, which is a residual of the process which determines polar motion from latitude observations. Polar motion is distinguished from azimuthal tilts at observing stations because polar motion has a particular geographic signature. Any anomalous tilts of the observing stations will register in the z term, along with errors in nutation, as well as refraction errors. The z term in the longest available polar motion data (the ILS data) has decade style variations of 10^{-6} rad or 10^3 times larger than the tilts that would be anticipated from the range of pressures implied by other geodetic data. It must therefore be dominated by nutation and refraction errors, and the possibility of extracting a tilt due to cmb pressure is remote.

Sea-level changes of about $2cm$ are also expected from the admissable changes in pressure at the core-mantle boundary. Changes in sea-level of about this magnitude are in principle detectable, but it is doubtful whether the cause could be separated from other influences, such as steric changes. Barnett[1983] reviews the difficulty in establishing the reality of a purported $10\,cm\,cy^{-1}$ increase in sea-level this century. The sea-level data do not exclude the possibility of decade-scale variations of about 2 cm.

Strains of 10^{-8} are expected. There are many strainmeters that can detect a strain of this magnitude, but it is doubtful they would have the necessary long term stability. Strains of this magnitude, and continental wavelength, imply trans-continental displacements of a few cm. This is detectable with long baseline interferometry.

Anticipated decade signals in gravity are about $.1\,\mu gal\,yr^{-1}$. This is much too small for absolute instruments to detect over a decade and, with the possible exception of the supeconducting gravimeter, it would be impossible to separate from the drift of a relative instrument.

Discussion

It has been shown that the decade variations in lod, polar motion and J_2 can all be used to set an upper bound on decade scale variations in the pressure field at the cmb. Of these, polar motion sets the smallest upper bound at $2\times10^3\,Pa$. By accounting for other sources of polar motion, such as the atmosphere and groundwater, it may be possible to reduce this even further. Utimately the possibility exists to use polar motion data to monitor the second degree tesseral harmonic of pressure and hence acquire some control on the large scale circulation near the cmb.

The decade variations in the lod set an upper limit of $3 \times 10^5\ Pa$ but it is likely that most of the decade variations in lod are from electromagnetic core-mantle coupling, so that these data do not contradict the upper bounds implied by the decade scale polar motion and J_2. Rather, they suggest that not more than about 10% of the decade variations in lod are due to pressure core-mantle coupling. As well, the decade changes in J_2, though poorly defined at present, imply decade variations in lod of at most $-11\ \mu sec\ yr^{-1}$ associated with changes in J_2. This is less than 10% of the decade changes in lod, further suggesting that no more than about 10% of the decade variations in lod might be due to pressure variations at the cmb.

Lageos data on ΔJ_2 will eventually prove to be more useful than the decade polar motion data, but for the present it is difficult to draw any firm conclusions because the data span is so short.

Acknowledgements. This work was supported by the Natural Sciences and Engineering Research Council of Canada, through operating grant A1084. The author would like to thank John Wahr, who pointed out an inconsistency in an earlier version of this work.

References

Barnett, T.P., Recent changes in sea level and their possible causes, Climatic Change, 5, 15-38, 1983.

Dickman, S.R., Investigation of some controversial polar motion features using homogeneous international latitude service data, Jour. Geophys. Res., 86, NO. B6, 4904-4912, 1981.

Le Mouel, J.L., C. Gire and T. Madden, 1985. Motions at the core surface in the geostrophic approximation, Phys. Earth Planet. Int., 39,270-280, 1985.

Markowitz, W., Latitude and longitude and the secular motion of the pole, in Methods and Techniques in Geophysics, 1, ed S.K. Runcorn, Interscience, N.Y., 325-361, 1960.

Merriam, J.B., Limits on lateral pressure gradients in the outer core from geodetic observations. Phys. Earth Planet. Int., 50, 280-290, 1988.

Morrison, L.V., Re-determination of the decade fluctuations in the rotation of the earth in the period 1861-1978, Geophys. Jour. Roy. Astron. Soc.,58, 349-360, 1979.

Nakiboglu, S.M. and K. Lambeck, Deglaciation effects on the rotation of the Earth, Geophys. Jour. Roy. Astron. Soc., 62, 49-58, 1980.

Peltier, W.R., Constraint on deep mantle viscosity from Lageos acceleration data, Nature, 304, 434-436, 1983.

Rubincam, D.P., 1984. Post-glacial rebound observed by Lageos and the effective viscosity of the lower mantle, Jour. Geophys. Res., 89, No. B2, 1077-1087, 1984.

Voorhies, C., Steady flows at the top of the Earth's core derived from geomagnetic field models, Jour. Geophys. Res., 91, B12, 12,444-12,467, 1986.

Wahr, J.M. The Earth's C_{21}, S_{21} gravity coefficients and the rotation of the core, Geophys. Jour. Roy. Astron. Soc., 88, No. 1 265-276, 1987.

Whaler, K.A., 1986. Geomagnetic evidence for fluid upwelling at the core-mantle boundary, Geophys. Jour. Roy. Astron. Soc., 86, No. 2, 563-588, 1986.

Whitworth, T., and R.G. Peterson, 1985. Volume transport of the antarctic circumpolar current from bottom pressure measurements. Jour. Phys. Ocean., 15, No. 6, 810-816, 1985.

Wilson, C.R. and R.O. Vicente, An analysis of the homogeneous ILS polar motion series, Geophys. Jour. Roy. Astron. Soc., 62, 605-616, 1980.

Yoder, C.E., J.G. Williams, J.O. Dickey, B.E. Schutz, R.J. Eanes and B.D. Tapley, Secular variation of the Earth's gravtiational harmonic J_2 from lageos and the non-tidal acceleration of the Earth's rotation, Nature, 303, 757-762, 1983.

Yumi, S. and K. Yokoyama, Results of the International Latitude Service in a homogeneous system 1899.9-1979.0, Pub. Cent. Bureau of the International Polar Motion Service, Mizusawa, 1980.

TIDAL FLOW IN THE EARTH'S CORE: A SEARCH FOR THE EPOCH AND AMPLITUDE OF EXACT RESONANCE IN THE PAST

Jacques Hinderer and Hilaire Legros

Laboratoire de Géodynamique, Institut de Physique du Globe
5, rue René Descartes 67084 Strasbourg Cedex, France

Abstract. We investigate the flow in the Earth's fluid core induced by the luni-solar tidal forces. When the frequency of some tidal waves is close to the nearly diurnal wobble eigenfrequency, a resonance process occurs. We intend to search for the epoch of exact resonance in the geological past and the resonant amplification with respect to present-day amplitude. Some geodynamical consequences are discussed.

Introduction

The precession and nutations of the Earth are caused by the luni-solar gravitational torques. In addition to this spatial motion, there is a tidally induced flow in the fluid core with respect to the mantle. The amplitude of this flow as a function of the tidal potential is dependent on several geometrical or geodynamical parameters relative to the Earth's model and points out the resonant behaviour in the vicinity of the nearly diurnal free wobble.
As the eigenfrequency of this rotational mode is related to the flattening of the core-mantle boundary, we search for the epoch of exact resonance in the past as a function of the Earth's rotation rate. We try then to set an upper bound on the resonant amplification factor when dissipation is taken into account. Some geophysical or geodynamical consequences are finally estimated.

Tidally Forced Response of an Earth's Model with a Fluid Core

The rotational response of the Earth and its fluid core to the luni-solar tidal gravitational potential is given by the Euler equations for conservation of angular momentum [e.g. Sasao et al., 1980; Hinderer et al., 1982]:

$$\dot{\omega}\left(1+\frac{\alpha k}{k_s}\right) - i\alpha\Omega\omega\left(1-\frac{k}{k_s}\right) + (\dot{\omega}^c + i\Omega\omega^c)\left(\frac{A^c}{A} + \frac{\alpha k_1}{k_s}\right) = \frac{3\alpha k}{a^2 k_s}\left(\frac{\dot{W}}{\Omega} + iW\right) - \frac{3i\alpha W}{a^2}$$

$$\dot{\omega}\left(1+\frac{q_\circ h^c}{2}\right) + \dot{\omega}^c\left(1+\frac{q_\circ h_1^c}{2}\right) + i\Omega\omega^c(1+\alpha^c + K' - iK) = \frac{3q_\circ h^c \dot{W}}{2a^2\Omega} \qquad (1)$$

These equations relative to the equatorial components of angular momentum are expressed here (in a complex notation) in the rotating Tisserand frame linked to the mantle; ω is the Earth's wobble, ω^c the core relative rotational motion with respect to the mantle (the flow is supposed, as usually in this kind of studies, to be of uniform vorticity) and Ω the Earth's uniform axial angular velocity. $\alpha = (C - A)/A$ is the Earth's dynamical ellipticity (C, A, polar and equatorial moment of inertia), α^c the flattening of the fluid core (actually the flattening of the core-mantle boundary in the case of a homogeneous core that is considered here) and $q_\circ = \Omega^2 a/g$ (a, Earth's radius; g, mean surface gravity) the classical geodynamical parameter expressing the ratio of centrifugal force to gravity. A^c is the core equatorial moment of inertia; k, k_1, h^c, h_1^c are elastic quantities (Love numbers or related combinations), k_s the secular Love number [Munk & MacDonald, 1960].
K, K' are dimensionless coupling constants of viscomagnetic origin.
Solving these equations as a function of the tidal potential W (here of degree 2 and order 1) leads, on one hand, to the amplitude of the well-known forced luni-solar precession-nutations (ω expresses actually the associated geographical motion of the axis of rotation) and, on the

other hand, to the amplitude of the core rotation ω^c with respect to the solid mantle. We consider hereafter only the tidal flow in the fluid core that can be written:

$$\omega^c = \frac{A(\alpha - \frac{q_o h^c}{2})\, 3W}{A^m(\sigma - \sigma_{nd})a^2} = f\, \frac{3\alpha W}{\alpha^c \Omega a^2} \quad (2)$$

where $A^m = A - A^c$ is the mantle moment of inertia. We have introduced a correcting factor noted f that involves the resonance process between some tidal waves of frequency σ close to the eigenfrequency σ_{nd} of the nearly diurnal free wobble (NDFW or free core nutation FCN) of expression:

$$\sigma_{nd} = -\Omega[1 + \frac{A}{A^m}(\alpha^c - \frac{q_o h_1^c}{2} + K' - iK)] \quad (3)$$

This eigenfrequency that appears in addition to the Chandlerian frequency in the presence of a fluid core is easily obtained by cancelling the right-hand side of equations (1).
In the peculiar case of the precessional excitation ($\sigma = -\Omega$), f becomes:

$$f = \frac{1 - \frac{q_o h^c}{2\alpha}}{1 - \frac{q_o h_1^c}{2\alpha^c}} \quad (4)$$

This factor is close to 0.5 for different Earth's models, and not 0.7 as previously supposed by analogy with the elastic reduction $(1 - k/k_s)$ of the Euler frequency. For a rigid mantle, the elastic quantities identically vanish and f = 1; in this case, mantle and core are precessing at the same rate, the core being simply tilted with respect to the mantle [Loper, 1975; Rochester, 1976].

Two very different situations can be investigated from equation (2): on one hand, the numerator of f may vanish, meaning that the core is motionless (with respect to the mantle) during tidal forcing; this can be due, for instance, to changes in the mantle mean rigidity that alters the value of the elastic quantity h^c, and is treated elsewhere [Hinderer et al., 1987]. On the other hand, the denominator of f may cancel ($\sigma - \sigma_{nd} \to 0$) leading to a resonance phenomenon in the tidal core response. Two major causes can be responsible for a resonant amplification: once again, a change in the mantle shear modulus (appearing by the parameter h_1^c) or a modification in the CMB flattening α^c. It can be shown that there is no physical solution (i.e. a positive value of shear modulus) for the precession, and that a very strongly rigid behaviour of the mantle (about 30 times the present one) is required to satisfy the resonance condition for the annual nutation (associated with the ψ_1 tidal wave).

Epoch of Exact Resonance

If the core-mantle boundary is supposed to be an equilibrium figure in response to the Earth's axial rotation, we can introduce a fluid Love number n_f in order to express the CMB flattening in the following manner:

$$\alpha^c = n_f(\frac{\Omega^2 a}{2gs}) \quad (5)$$

where s = b/a is the ratio of core to Earth radii.
The resonance condition is then:

$$\sigma - \sigma_{nd} = \sigma + \Omega + \Omega^3 \frac{Aa}{2A^m gs} n_f (1 - \frac{s h_1^c}{n_f}) = 0 \quad (6)$$

We immediately see that there is no solution for the precession ($\Omega \to 0$). On the contrary, if we suppose that the orbital motion of the Earth around the Sun has been stable in the geological past [e.g. Toomre, 1974], the condition for exact resonance for the retrograde annual nutation($\sigma = -\Omega^*(1 + 1/365)$) is:

$$\frac{\Omega^*}{365} = B\,\Omega^3 \quad (7)$$

noting Ω^* the present value of the angular velocity and B the coefficient of Ω^3 in (6); the constancy of B implies that the mantle elastic (h_1^c) and fluid (n_f) behaviours have not significantly changed in the past, in addition to a stable core size (s) and constant mean densities of the core and shell (A/A^m).
Solving (7) leads to the value of the past angular velocity Ω and, using a given value of the secular rate of deceleration $d\Omega/dt$, to the epoch of resonance T. The mean estimated value of $d\Omega/dt$ is $-6.4\ 10^{-22}$ rd.s^{-2} [Stephenson & Morrison, 1982], but some larger values are sometimes invoked; we take the value $d\Omega/dt = -11.6\ 10^{-22}$ rd.s^{-2} [Cannon, 1974] as an upper bound.

If we start from the theoretical value of the nearly diurnal eigenfrequency $\sigma_{nd} \approx -\Omega^*(1 + 1/465)$ [Sasao et al., 1980; Wahr, 1981], we have then $B(\Omega^*)^2 = 1/465$, and the condition $(\Omega/\Omega^*) = (465/365)^{1/3}$ leads to:

$$170 \text{ Ma} \le T \le 300 \text{ Ma} \quad (8)$$

according to the two previous rates of deceleration; this range includes the value (200 Ma) already proposed [Toomre, 1974].

If we start from the value inferred from recent observations in VLBI measurements [Gwinn et al., 1986] or gravity tides in Central Europe [Neuberg et al., 1987], $\sigma_{nd} \approx -\Omega^*(1 + 1/435)$ and the epoch of resonance becomes:

$$120 \text{ Ma} \le T \le 220 \text{ Ma} \qquad (9)$$

Notice the proximity (may be simply coincidental) of T with the onset of continental drift (≈ 180-200 Ma) and with one of the major biological extinction in the late Permian (≈ 245 Ma).

If the discrepancy between theory and observation is attributed to a non-hydrostatic CMB flattening value in σ_{nd}, the search for the epoch of resonance is, in general, more complicated (there is an additional term proportional to Ω in equation (6)); it can however be shown that, even in this case, the results are very similar.

Amplitude of Exact Resonance

For a conservative system, i.e. when there is no viscomagnetic friction at the CMB and when the mantle rheology is perfectly elastic, the amplification of the core response would be infinite at exact resonance. But, as dissipation is present both in the core-mantle coupling and in the mantle deformation, the resonance will be of course finite. The correcting parameter f involves then complex Love number combinations ($h^c \to \tilde{h}^c$, $h_1^c \to \tilde{h}_1^c$) and an imaginary contribution to the eigenfrequency (damping proportional to the coupling constant K) and becomes also a complex quantity in equation (2). It follows that, at exact resonance ($\sigma - \mathrm{Re}(\sigma_{nd}) \to 0$), the amplitude of the core flow is then (here in terms of friction only):

$$\left| \omega_r^c \right| \approx \frac{1 - \frac{q_\circ h^c}{2\alpha}}{K} \frac{3\alpha W}{\Omega a^2} \qquad (10)$$

Therefore, any attempt to estimate the amplitude of resonant amplification needs the extrapolation of dissipative mechanisms into the past, and this is certainly not obvious. We have however tried to quantify this effect simply by supposing that no fundamental change has occured in these mechanisms over the last 200 Ma. Taking into account only CMB friction leads to a ratio of resonant to present amplitude of about $4.\ 10^3$, and a similar calculation relative to mantle anelasticity using recent (theoretical) estimates of the damping of the nearly diurnal eigenmode [Wahr & Bergen, 1986] yields a ratio of about $4.\ 10^2$. We see that even weakly dissipative processes in the Earth's interior strongly limit the resonance.

Some Consequences

The consequences relative to this nearly diurnal resonance in the tidal flow are of several kinds.First, the core motions are amplified by the same ratio as given above, and the velocity field at the CMB can reach an amplitude of the order of $3.\ 10^{-1}$ $m.s^{-1}$, which is quite large in the context of core dynamics; remember that a typical fluid velocity (as the one related to the westward drift of the Earth's magnetic field) is rather of order 10^{-4} $m.s^{-1}$.

If friction stresses exist at the CMB, some power can be extracted from the rotational energy and converted into heat [e.g. Stacey, 1973]; using the previous velocity amplitude leads to a power of about 10^{11} Watt that is actually only slightly smaller than the dissipation of the whole Earth's rotational energy related to the secular tidal deceleration [Lambeck, 1975a], but about 1600 times the power dissipated by the same mechanism during the precessional motion of the core.

The fluid pressure associated with the resonant velocity is found to be of the order of 10^6 Pascal (10 bar) and leads to an enhanced mantle deformation with surface displacements of a few meters and stresses in the lithosphere of a few bars. This deformation is not negligible but, by no means, exceptional (it is simply ≈ 30 times larger than the present-day tidal deformation associated with the precessional main wave K_1).

In addition to these mechanical effects, we can think of thermal consequences related to the heat release near the CMB that can be associated to convection in the core (with implications for the generation of the Earth's magnetic field) and, besides, that could trigger upwelling of hot material (plume) from the lower mantle to the lithosphere [e.g. Loper & Stacey, 1983] and lead to volcanism.

Finally, let us mention that a very similar resonance process exists between the annual atmospheric forcing (seasonal redistribution of masses) and the well-known Chandler wobble [Cannon, 1974; Lambeck, 1975b], that is the second rotational eigenmode deriving from Euler equations (1).

As a concluding remark, we think that this nearly diurnal resonance process in the fluid core is not of negligible geophysical significance, especially if there is a double resonance in the past with coincidental epochs of resonance and cumulative energetic consequences.

Acknowledgments. This study was supported by INSU, France (ATP 'Noyau' grant 511614).

References

Cannon, W.H., 1974. The Chandler annual resonance and its possible geophysical

significance, Phys. Earth planet. Int., 9, 83-90.
Gwinn, C.R., Herring, T.A. & Shapiro, I.I., 1986. Geodesy by radio interferometry: Studies of the forced nutations of the Earth. 2. Interpretation, J. Geophys. Res., 91, 4755-4765.
Hinderer, J., Legros, H. & Amalvict, M., 1982. A search for Chandler and nearly diurnal free wobbles using Liouville equations, Geophys. J. R. astr. Soc., 71, 303-332.
Hinderer, J., Legros, H. & Amalvict, M., 1987. Tidal motions within the Earth's fluid core: resonance process and possible variations, Phys. Earth planet. Int., 49, 213-221.
Lambeck, K., 1975a. Effects of tidal dissipation in the oceans on the Moon's orbit and the Earth's rotation, J. Geophys. Res., 80, 2917-2925.
Lambeck, K., 1975b. The Chandler annual resonance, Phys. Earth planet. Int., 11, 166-168.
Loper, D.E., 1975. Torque balance and energy budget for the precessional driven dynamo, Phys. Earth planet. Int., 11, 43-60.
Loper, D.E. & Stacey, F.D., 1983. The dynamical and thermal structure of deep mantle plumes, Phys. Earth planet. Int., 33, 304-317.
Munk, W.H. & MacDonald, G.J.F., 1960. The Rotation of the Earth, Cambridge University Press, 323 pp.
Neuberg, J., Hinderer, J. & Zürn, W., 1987. Stacking gravity tide observations in Central Europe for the retrieval of the complex eigenfrequency of the nearly diurnal free wobble, Geophys. J. R. astr. Soc., 91, 853-868.
Rochester, M.G., 1976. The secular decrease of obliquity due to dissipative core-mantle coupling, Geophys. J. R. astr. Soc., 46, 109-126.
Sasao, T., Okubo, S. & Saito, M., 1980. A simple theory on dynamical effects of a stratified fluid core upon nutational motion of the Earth, in Proc. IAU Symp. No 78, Nutation and the Earth's rotation, Kiev, May 1977, E.P. Fedorov, M.L. Smith & P.L. Bender eds., D. Reidel, Dordrecht, 165-183.
Stacey, F.D., 1973. The coupling of the core to the precession of the Earth, Geophys. J. R. astr. Soc., 33, 47-55.
Stephenson, F.R. & Morrison, L.V., 1982. History of the Earth's rotation since 700 B.C., in Tidal friction and the Earth's rotation, II, P. Brosche & J. Sundermann eds., Springer-Verlag, Berlin, 29-50.
Toomre, A., 1974. On the 'nearly diurnal wobble' of the Earth, Geophys. J. R. astr. Soc., 38, 335-348.
Wahr, J.M., 1981. Body tides of an elliptical, rotating, elastic and oceanless earth, Geophys. J. R. astr. Soc., 64, 677-703.
Wahr, J. M. & Bergen, Z., 1986. The effects of mantle anelasticity on nutations, earth tides, and tidal variations in rotation rate, Geophys. J. R. astr. Soc., 87, 633-668.

CRYSTALLIZATION OF THE EARTH'S INNER CORE

Orson L. Anderson

Institute of Geophysics and Planetary Physics, UCLA, Los Angeles, California 90024

David A. Young

Lawrence Livermore National Laboratory, University of California, Livermore, California 94550

Abstract. The temperature and the volume change across the ε-iron solid-liquid transition are now known from two different calculations: the first is an ab initio calculation by Young and Grover [1984], and the second is from the construction of the P-T-V phase diagram [Anderson, 1986] constrained by the shock-wave data of Brown and McQueen [1982, 1986]. These calculations lead to the heat of crystallization of pure iron at the pressure of the Earth's inner-outer core boundary being

$$200 \text{ cal/g} < \Delta H_m < 375 \text{ cal/g}.$$

Considering the freezing point depression of probable impurities in the inner core, we find the heat flow into the mantle about twice that of the latest estimate made by Verhoogen [1980].

Introduction

In 1961, Verhoogen introduced the concept that an important source of energy production in the core is the heat of crystallization as the inner solid core crystallizes from the liquid. This is consistent with a cooling planet cosmology. The important parameter is ΔH_m of iron, the change in enthalpy at melting, evaluated at 330 Gpa, the present pressure at the inner-outer core boundary, modified to account for the solute impurities. Using the best estimates of $(dT/dP)_m$ and ΔS_m available from the then current theories of melting, he found ΔH_m = 65 cal/gm for iron, and revised it upward to 106 cal/gm [Verhoogen, 1980].

Because of new ab initio calculations on the melting of iron, and because of the important shock wave results of Brown and McQueen [1982, 1986], it is possible more carefully to limit the melting temperature, T_m, of iron at P = 330 Gpa, and place some limits on ΔV_m. Anderson [1982, 1986] found a solution to the phase diagram of iron which incorporated the shock wave work of Brown and McQueen, and thus a more accurate Combining the effects arising from Eqs. (2) and (3), we can estimate the change in the heat of crystallization arising from the impurities sulfur and oxygen.

Effect of Impurities on the Heat of Crystallization, ΔH_m

Using Eqs. (1) and (2), along with the criterion that $\Delta G = 0$, or thermal equilibrium, between the solid pure phase and the liquid with solutes, whence $\Delta H_m = T_m \Delta S_m$, one can calculate the effect of impurities upon ΔH_m. We consider approximations for the case of oxygen [Ringwood, 1977] and sulfur [Stevenson, 1980].

Jeanloz and Ahrens [1980] presented shock wave Hugoniot data on Fe and FeO, from which one can calculate the oxygen atomic concentration which satisifies the pressure and density data of the outer core. About 45 wt percent FeO is required in combination with 55 wt percent pure iron to satisfy the outer core data at 250 GPa. From these data the value of χ_s is about 0.3, which satisfies the measured density of the inner core between 150 and 250 GPa (see Figure 1).

Using data in a paper by Ahrens [1979], who presented shock wave data on FeS_2 and $Fe_{0.9}S$ it is easily shown that the sulfur/iron atomic ratio is about 0.096, so that χ_s is close to 0.083. Using equation (3) we have

$$\text{oxygen:} \quad \Delta S = \Delta S_m - 8.314 \times 10^7 \ \ell n(1 - 0.3) \text{ erg/mole} = \Delta S_m + 2.97 \text{ J/mole},$$

$$\text{sulfur:} \quad \Delta S = \Delta S_m - 8.314 \times 10^7 \ \ell n(1 - 0.083) \text{ erg/mole} = \Delta S_m + 0.72 \text{ J/mole}.$$

Heat of Crystallization of Pure Iron

Young and Grover [1984] calculated the liquid phase boundary for pure ε (hcp) iron, and found T

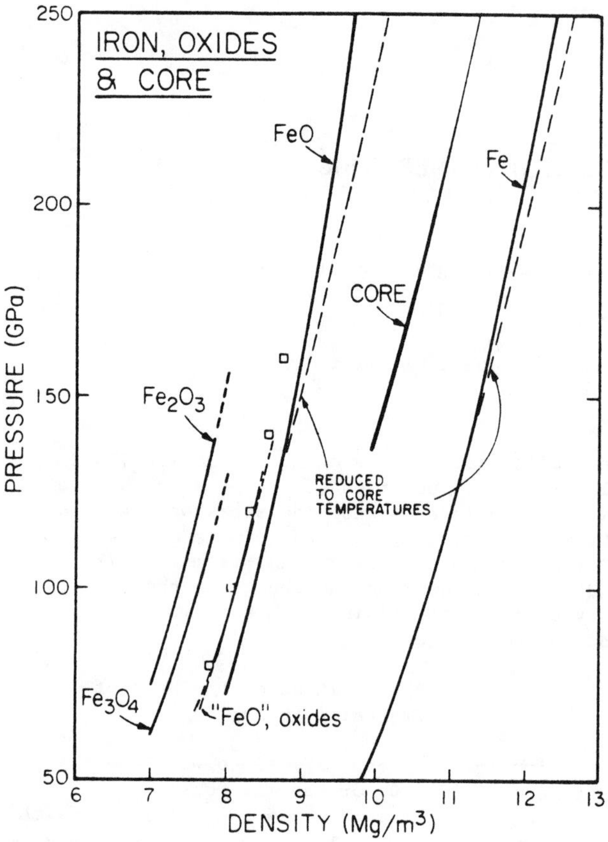

Fig. 1. Hugoniot curves for FeO, Fe, and the Earth's outer core. Published by Jeanloz and Ahrens [1980].

and ΔS at P = 330 GPa to be 6600 K and 8.16 × 10^7 erg/mole K. For the pure iron, then, ΔH(P=330) = TΔS = 5.346 × 10^{11} erg/mole = 229 cal/gm using their theoretical calculations.

Anderson [1986] constructed the phase diagram of pure iron, constrained by the shock wave data of Brown and McQueen [1980, 1982, 1986]. He found a triple point at 280 GPa (see Figures 2 and 3). Above the triple point he found ΔV_m = 0.14 cc/mole and the slope $\Delta T_m/\Delta P$ = 10.2° GPa^{-1}, giving ΔS = 13.98 × 10^7 erg/mole. This is larger than the ΔS found by Young and Grover, because ΔV above the triple point at 280 GPa combines ΔV of method of estimating ΔH_m is now within reach.

In this paper, we re-evaluate ΔH_m, finding it between three and five times greater than in Verhoogen's first estimate and more than twice as much as his second estimate. This substantially increases the calculated heat production in the core from the heat of fusion.

Temperature Depression by Impurities

If the impurity content is a few percent, the melting curve of an iron mixture at 330 Gpa is depressed from that of pure iron by approximately several hundreds of degrees, perhaps as much as 1000 K. We wish to indicate how this figure is arrived at. Supposing that the majority of the impurities in the iron core are in the outer core, then let us consider thermodynamic equilibrium at the inner-outer core boundary pressure between solid pure iron and liquid iron with solute.

Equilibrium is defined by the following equation appropriate to ideal mixing, when only one species of impurity is considered:

$$\mu_o^{\ell} + KT\ell n(1 - \chi_s) = \mu_o^{s}(P,T) \qquad (1)$$

where μ_o^{ℓ} and μ_o^{s} are the chemical potentials of liquid and solid iron, and χ_s is the concentration of solute in the liquid phase. From this relationship, to a second approximation, the depression in the melting point is given by [Atkins, 1986]:

$$\ell n(1 - \chi_s) = -\left(\frac{\Delta H_m}{R}\right)\left(\frac{1}{T} - \frac{1}{T^*}\right) = -\left(\frac{\Delta S_m}{R}\right)\left(1 - \frac{T}{T^*}\right), \qquad (2)$$

when T* is the melting temperature of the pure substance at χ_s = 0. From these formulas we can estimate that a few percent of an impurity in

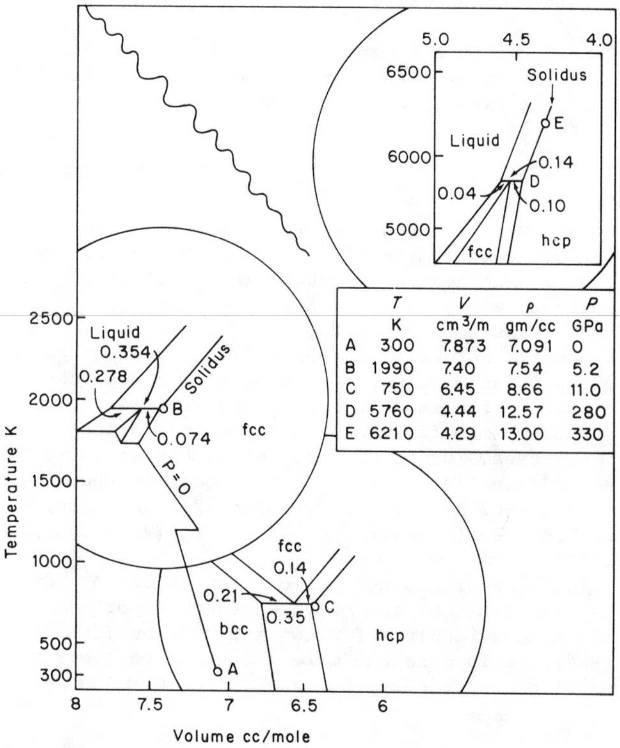

Fig. 2. The volume temperature phase diagram of iron, showing ΔV across the phase boundaries for the three triple points. Published by Anderson [1986].

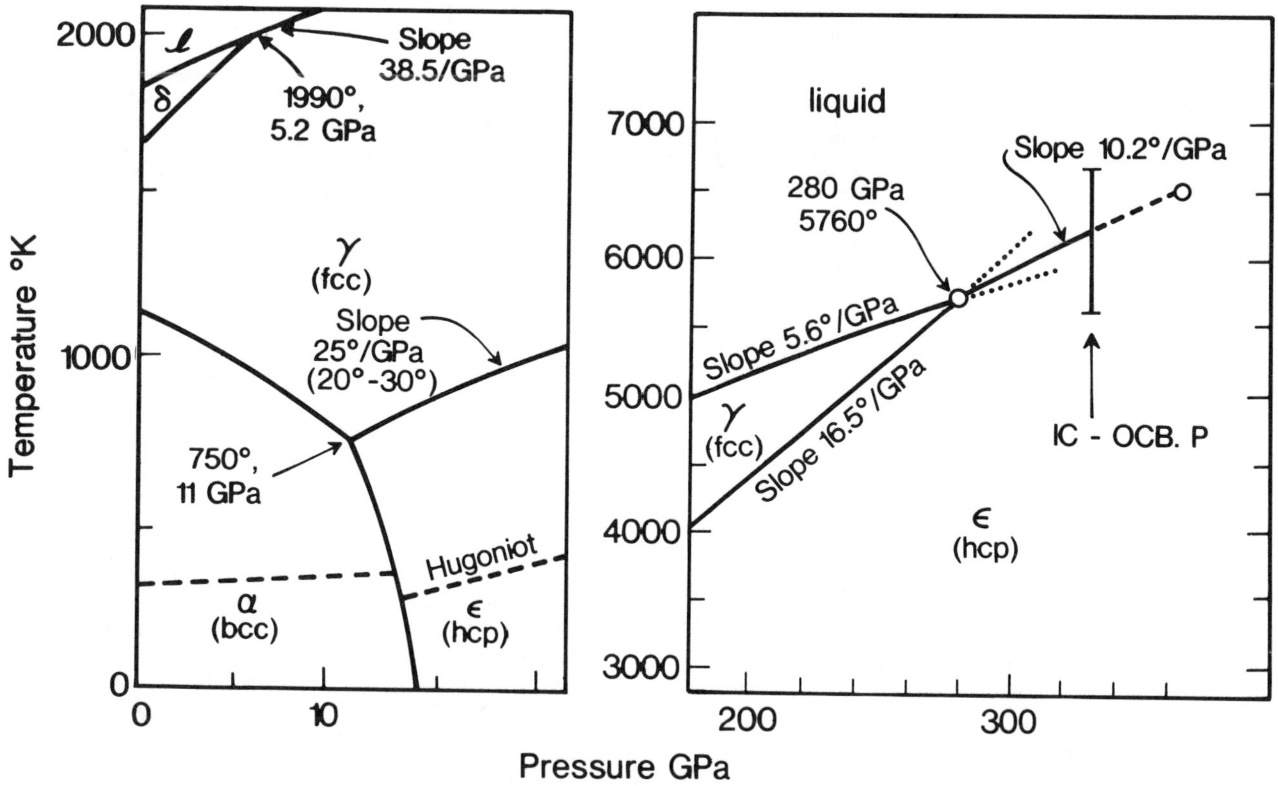

Fig. 3. The pressure-temperature phase diagram of iron. (a) modified from Liu [1975], indicates the tp's at 5.2 GPa as determined by Mao et al. [1967] and at 11 GPa as determined by Strong et al. [1973], and Bundy [1965]. Published by Anderson [1986].

iron can drop the melting point by 800-1000 C, if T* is of the order of 6000 K.

Entropy of Mixing

An entropy of mixing term increases when the impurity concentration increases, and this will tend to increase the heat of crystallization, compensating in part at least for the lowering of ΔH_m by freezing point depression. The entropy change at melting due to changing the impurity of the liquid by χ_s is

$$\Delta S = \Delta S_m - R\ell n(1 - \chi_s) . \qquad (3)$$

the two lower branches (see Figures 2 and 4). However, the temperature at P = 330 GPa is lower than found by Young and Grover, being 6210 K according to Anderson [1986]. The heat of crystallization, according to Anderson, is therefore

$$\Delta H_m = T_m \Delta V_m \left(\frac{\Delta P}{\Delta T_m}\right) = 372 \text{ cal/gm}. \qquad (4)$$

Using equation (2), we find for oxygen that ΔT = 2402 K from the Young and Grover [1984] data, and ΔT = 1290 K from the Anderson [1986] data.

For sulfur, ΔT = 587 according to Young and Grover's data, and ΔT = 321 according Anderson's data.

The decrease in ΔH due to freezing point depression is calculated by finding $(RT/\Delta S_m)\ell n(1 - \chi_s)\Delta S_m$. The entropy of mixing is $-RT\ell n(1 - \chi_s)$ so that the value of ΔH including mixing is

$$\Delta H_m(\text{mix}) = T_o \Delta S_m - \frac{R^2 T_o}{\Delta S_m}\left[\ell n(1 - \chi_s)\right]^2 . \qquad (5)$$

Using equation (5) and the Young and Grover [1984] data, we find ΔH_m for the impure liquid is 200 cal/gm for oxygen impurities and 226 cal/gm for sulfur, compared with 229 cal/gm for the pure iron. From the Anderson [1986] data we find ΔH_m for the impure liquid to be 355 cal/gm for oxygen impurities and 370 cal/gm for sulfur, compared with 372 for pure iron.

We see, therefore, that for sulfur the effect of impurities in the liquid iron of the outer core has little effect on the heat of crystallization at the inner-outer core boundary and can be ignored. For the case of oxygen impurities, there appears to be a slight effect. We anticipate that oxygen would lower the heat of crystal-

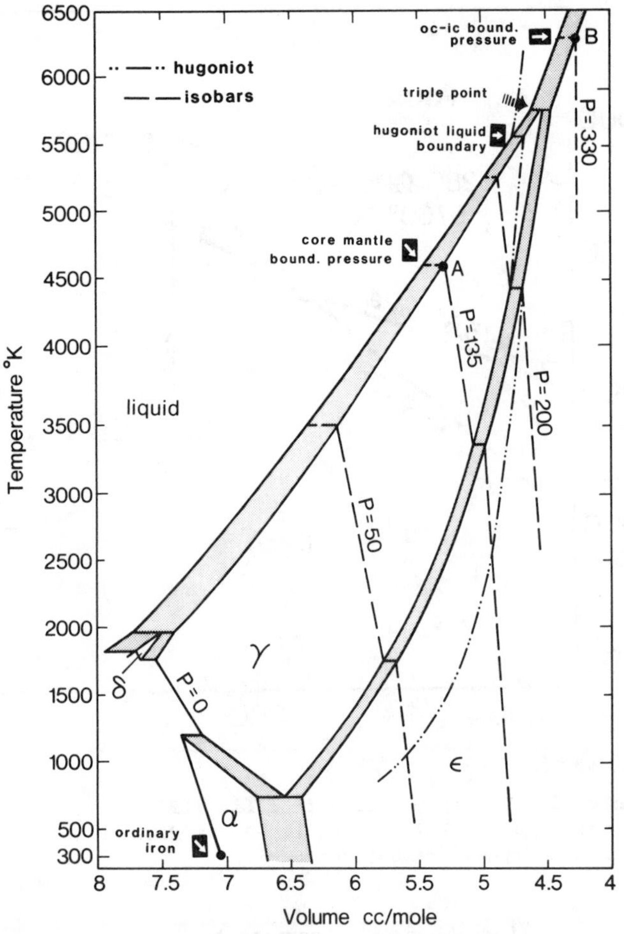

Fig. 4. The volume-temperature phase diagram of iron. Hatched regions show the ΔV due to structural transitions and fusion. A and B represent the limits in pressure found in the outer core of the Earth. α, ε, and γ represent the three main structural phases of iron. The phase diagram is determined primarily [Anderson, 1986] from the Hugoniot's intersection with the ε-γ transition and the γ-liquid transition found by Brown and McQueen [1982, 1986]. Note that ΔV_m increases above the triple point temperature.

lization by about 6%, but the formulas are suspect because of the high concentration of solute. We therefore estimate that the heat of crystallization at the core's boundary is bounded by

$$200 \text{ cal/gm} < \Delta H_m < 375 \text{ cal/gm}. \qquad (6)$$

considering the Young and Grover [1984] model and the Anderson [1986] model as extrema. The estimate given by equation (6) is close to the value of 10^6 J/Kg used in the calculations of Gubbins et al. [1979] and those of Mollett [1984].

Purification of the Inner Core by Equilibrium Freezing

Let us assume the cooling Earth paradigm and examine the consequences of a slowly cooling core. Verhoogen [1961] proposed that slow cooling and crystallization of the core, accompanied by a corresponding growth of the inner core, provides much of the heat necessary for convection in the outer core. This is analogous to metallurgical processing with fractional crystallization by a moving solid-liquid interface. In any solid where the solute depresses the freezing point, and the solidus is distinguishable from the liquidus, the solute concentration in the freezing solid is lower than that of the liquid [Pfann, 1958]. A slowly moving inner core boundary (ca. 1 cm/year) is analogous to a long bar undergoing equilibrium freezing. Here the solute concentration in the solid is k_o times that of the liquid, where k_o is the equilibrium distribution coefficient (k_o < 1) of the solute.

Thus, Verhoogen's idea results in concluding that the solute concentration in the liquid is many times more than in the solid, the ratio depending upon k_o at core pressures. A purified inner core growing at the expense of the outer core would result in a marked compositional difference between the inner and the outer cores [Masters, 1979] and would provide a major source of the Earth's heat flow.

In the zone melting process of purification, it is found that normal freezing can be a very efficient method of purification. A slowly moving inner-outer core boundary could leave iron in a more or less pure solid state, all depending, of course, on the magnitude of the equilibrium distribution coefficient of the impurities. While we do not know how pressure affects the equilibrium distribution coefficient of sulfur, we know it is small at room pressure, and it may be even smaller at core conditions.

As the core grows and cools over geologic time, the impurity content increases in the outer core. The impurities are released at the inner core-outer core boundary and, being less dense than iron, tend to rise. This releases gravitational energy. The solidification process also releases latent heat. These two energy releases provide not only a heat source for the mantle but also a mechanism of continuous mixing of the outer core, thus tending to make it homogeneous. The mixing of the lighter elements as they rise toward the core-mantle boundary is available to help generate the Earth's magnetic field.

If heat is released at the crystallization interface, then the actual thermal gradient may exceed the adiabatic gradient; i.e., there is a superadiabatic component to the temperature. This is a condition of convection, but the superadiabaticity need not be great if the velocity of convection is not large. If the superadiabaticity were large, then there would not be a substantial vertical motion in the flow, which is required by the dynamo [Busse, 1981]. Since growth of the inner core provides both compositional and thermal destabilization [Fern and Loper, 1981], stable regions and stratification should not develop in the outer core, agreeing with the index of homogeneity computed there from seismic data.

A growing core satisfies the requirement for generating a magnetic field, and internal heating by such radioactive heat sources as K^{40} is not required [Stevenson, 1981; Stacey, 1980].

Energy Arising from Crystallization of the Inner Core

If the average temperature of the whole core decreases by ΔT, crystallization will occur at the inner core boundary, which will move outwards by a distance Δr. The thermal gradient at the inner core-outer core boundary, taken from data on the solid side, is [Anderson, 1986]

$$\Delta T/\Delta r = \frac{6210 - 6450}{1833} = -0.131 \text{ deg/km} = \lambda. \quad (7)$$

The total amount of heat released by a drop in ΔT is [Verhoogen, 1961]

$$\Delta Q_c = \left(\frac{C_P M_c + 4\pi r_c{}^2 \rho \Delta H_m}{\lambda}\right)\Delta T \quad (8)$$

$$= \Delta Q_1 + \Delta Q_2 . \quad (9)$$

The first term is heat loss by cooling alone and the second arises from the heat released by crystallization. M_c is the mass of the whole core, r_c is the radius of the inner core, and ΔH_m is the heat of fusion.

Now, calculating the first term in equation (9),

$$\Delta Q_1 = C_P M_c \Delta T. \quad (10)$$

Taking C_p = 3.9k/atom for the liquid at the inner-outer core boundary [Stevenson, 1981] (7.8 cal/deg/mole × 55 gm/mole), then C_p = 0.14 cal/gm/deg. There being 1.92 × 10^{27} gm in the core, we have from equation (10)

$$\frac{\Delta Q_1}{\Delta T} = 2.7 \times 10^{26} \text{ cal/deg.} \quad (11)$$

Let us calculate the second term in equation (9):

$$\frac{\Delta Q_2}{\Delta T} = \frac{4\pi r_c{}^2 \rho \Delta H_m}{\lambda}, \quad (12)$$

where ρ = 13 gm/cc, r_c = 1.183 × 10^8 cm, ΔH_m = 200 to 375 cal/gm, and λ = 1.31 × 10^{-6} deg/cm. Thus, for pure solid iron in equilibrium with liquid iron containing a light solute of oxygen or sulfur, the cooling heat loss due to crystallization is, from equation (12)

$$\frac{\Delta Q_2}{\Delta T} = 3.3 \text{ to } 6.3 \times 10^{26} \text{ cal/deg,} \quad (13)$$

and the total cooling loss is, from equations (11) and (12)

$$\frac{\Delta Q_c}{\Delta T} = 6 \text{ to } 9 \times 10^{26} \text{ cal/deg.} \quad (14)$$

The Heat Flow Due to the Cooling Core

The calculation of the heat flow involves the rate of growth of the inner core, so an assumption has to be made about the lapsed time since the initiation of crystallization. One could assume, for example, that the inner core began to form and started to grow at the earliest time in the Earth's history, in which case the lapsed time would be 4.5 billion years. We shall see that the temperature drop of the core due to crystallization is at most a few hundred degrees, so it seems unlikely that the temperature was fortuitously exactly at the fusion point the moment the liquid core condensed. The initiation of crystallization was very likely delayed, but it is difficult to verify the lapsed time. Verhoogen [1961] used 3 billion years and Loper [1978] used 4 billion years for lapsed time.

It is clear from the study of magnetization of old rocks that the Earth has possessed a magnetic field for at least 3 billion years, and that the dipole movement has been fairly constant [Gubbins et al., 1979]. Calculations using τ from 3 to 4 billion years appear reasonable.

The rate of energy production can be found from

$$\frac{dQ_c}{dt} = \left(\frac{dQ_c}{dT}\right)\left(\frac{dT}{dr}\right)\left(\frac{dr}{dt}\right). \quad (15)$$

The thermal gradient dT/dr will surely vary with time, but for this calculation we shall use the present value given by equation (7). Using equations (14) and (7), equation (15) becomes

$$\frac{\Delta Q_c}{\Delta t} \simeq \frac{18 \times 10^{12}}{\tau} \text{ Watts} \quad (16)$$

where τ is lapsed time in billions of years.

To calculate the rate of cooling of the core, use equations (14) and (16) in the following:

$$\frac{\Delta T}{\Delta t} = \left(\frac{\Delta Q_c}{\Delta t}\right)\left(\frac{\Delta Q_c}{\Delta T}\right) = \frac{43.47}{\tau} \text{ deg/Ga}$$

(deg per billion years) (17)

This agrees well with the estimated value of 10 deg/Ga used by Gubbins et al., [1979] in three of their models.

The rate of cooling is not constant since T, ρ, λ, and r_c depend upon time, but we can get the approximate value of ΔT during time τ (see Table 1). Our results for ΔQ_c/dT are a little higher than those obtained by Gubbins et al. [1979], but not significantly so. We see that for τ = 3 the heat generated by cooling, including crystallization, as shown in the Table 1, is a little over 0.5 × 10^{13} W and about equal to the estimate of adiabatic conductive heat in the outer core.

TABLE 1. Heat Flow and Heat Production from Crystallization of the Inner Core

Time of Core Crystallization τ (billion years)	Heat Flow $\Delta Q_c/\Delta T$ (10^{12} Watts)	Total Temperature Drop (degrees)
4.5	3.8	196
4	4.3	174
3.5	5	152
3	5.75	131
2.5	6.9	109

If the core has been steadily crystallizing for the last 4 billion years, then the average rate of advance of the inner-outer core boundary, from thermal contraction alone, is

$$\frac{dr_c'}{dt} = \frac{1.185 \times 10^8 \text{ cm}}{4 \times 10^9 \text{ year}}$$

$$= 0.03 \text{ cm/year} = .95 \times 10^{-9} \text{ cm/sec},$$

a very slow rate indeed allowing thermal equilibrium to be maintained.

Gravitational Energy Arising from Zone Refining the Core

If the value of the distribution coefficient is small enough, as it most likely is for sulfur in iron at core conditions, then the crystallization of the core boundary will eject a lighter constituent into the liquid core. This is demonstrated by the density jump at the inner core (see, for example, Bolt and Uhrhammer [1981]). Thus a growing core will gradually increase the density distribution of the core towards a more dense center, and the resulting increase in gravitational energy will result in heat release. Central condensation of iron in the core causes g to rise and so the pressure also rises. The total radius of the Earth also changes. The detailed theory of gravitational energy release by a light constituent ejection into the outer core from the inner core was made by Loper [1978]. He assumed a two-phase core composed initially of iron and a lighter constituent in which the mass fraction of the lighter constituent was 0.087 prior to the beginning of core crystallization. He calculated that the gain of gravitational energy due to core growth (ejection of lighter impurities) is

$$\left(\frac{dQ_c}{dT}\right)_{g.r.} = 2.2 \times 10^{12} \text{ W} \qquad (18)$$

for a core growth period of 4 billion years. (Loper found a total gravitational release of 2.5 × 10^{29} J over the total lapsed time since crystallization. For τ = 3, the heat flow from this effect would be close to 3 × 10^{12}.)

Thus, the total output due to core growth resulting from cooling (including crystallization) and zone refining is close to 8 × 10^{12} W. This is quite sufficient to run the dynamo if the thermal efficiency is a few percent. For example, Gubbins et al. [1979] required 5 x 10^{11} W for an average torrodial field of 240 gauss. Mollett [1984] found that the dynamo requirements for heat flux from core including gravitation were of 2 to 7 × 10^{12} W, and he concluded that the heat flux out of the core can be sufficiently large to drive a dynamo at the present time and at times previous to the formation of the core. On the basis of our calculations, we see that total heat flux from the core is almost but not quite enough to supply the required energy balance of heat flow at the Earth's surface [Sclater et al., 1980]. Thus we conclude that if K^{40} is present in the Earth it may only be a low-potassium model and not the high concentration model suggested by Goettel [1976]. If there is a small concentration of uranium and thorium in the lower mantle then the heat balance at the Earth's surface is satisfied and no potassium is needed in the core at all; or if the lapsed time of crystallization is close to 2 billion years, then no K^{40} is needed.

According to Loper's model calculation, by the time the inner core grows to its present size, the concentration of the lighter constituent in the outer core has grown from a mass fraction of 0.08 to 0.0928, which means that a corresponding amount of iron has been taken from liquid and concentrated in the inner core.

Acknowledgments. This research was performed under the auspices of the U.S. Department of Energy by the Lawrence Livermore National Laboratory under Contract No. W-7405-ENG-48. IGPP Contribution No. 3100.

Note added in proof. Recent experiments with the diamond cell indicate that the melting temperature of iron may be as high as 7000 K [Williams et al., 1987] at the inner-outer core boundary pressure, in contrast to the report here of 6210 K. A higher temperature T_m would not substantially change ΔH_m, because of the rigid constraints on the melting of iron below 20 GPa. A higher T_m would result in a smaller ΔV_m, and therefore a smaller ΔS_m, so that compensation would tend to take place in ΔH_m. The effect would be as if Figure 4 underwent a topographical stretch along the T axis.

References

Ahrens, T.J., Equations of state of iron sulfide and constraints on the sulfur content of the Earth, *J. Geophys. Res.*, *84*, 985-998, 1979.

Anderson, O.L., The Earth's core and the phase diagram of iron, *Philos. Trans. R. Soc. London, A*, *306*, 21-35, 1982.

Anderson, O.L., Properties of iron at the Earth's core conditions, *Geophys. J. R. Astron. Soc.*, *84*, 561-580, 1986.

Atkins, P.W., *Physical Chemistry*, pp. 217-219, W.H. Freeman and Co., San Francisco, 1986.

Bolt, B., and R.A. Uhrhammer, The structure, density, and homogeneity of the Earth's core, in *Evolution of the Earth*, edited by R.J. O'Connell and W.S. Fyfe, pp. 28-29, American Geophysical Union, Washington, D.C., 1981.

Brown, J.M., and R.G. McQueen, Melting of iron under core conditions, *Geophys. Res. Lett.*, *7*, 533-536, 1980.

Brown, J.M., and R.G. McQueen, The equation of state for iron and the Earth's core, in *High Pressure Research in Geophysics*, edited by A. Akimoto and M. Manghnani, pp. 611-625, Center for Academic Publication, Tokyo, 1982.

Brown, J.M., and R.G. McQueen, Phase transitions, Grüneisen parameter and elasticity for shocked iron between 77 Gpa and 400 Gpa, *J. Geophys. Res.*, *91*, 7485-7494, 1986.

Bundy, F.P., Pressure-temperature phase diagram of iron to 200 kbar, 900°C, *J. Appl. Phys.*, *36*, 616-620, 1965.

Busse, F.H., Dynamics of the Earth's core and the geodynamo, in *Evolution of the Earth*, edited by R.J. O'Connell and W.S. Fyfe, pp. 53-58, American Geophysical Union, Washington, D.C., 1981.

Fern, D.R., and D.E. Loper, Compositional convection and stratification of the Earth's core, *Nature (London)*, *289*, 393-394, 1981.

Goettel, K.A., Models for the origin and composition of the Earth, and the hypothesis of potassium in the Earth's core, *Geophys. Surv.*, *2*, 369-397, 1976.

Gubbins, D., T.G. Masters, and J.A. Jacobs, Thermal evolution of the Earth's core, *Geophys. J. R. Astron. Soc.*, *59*, 57-99, 1979.

Jeanloz, R., and T.J. Ahrens, Equations of state of FeO and CaO, *Geophys. J. R. Astron. Soc.*, *62*, 505-528, 1980.

Liu, L. On the (γ, ε, ℓ) triple point of iron and the Earth's core, *Geophys. J. R. Astron. Soc.*, *43*, 697-705, 1975.

Loper, D.E., The gravitationally powered dynamo, *Geophys. J. R. Astron. Soc.*, *54*, 389-404, 1978.

Mao, H.K., Bassett, W.A., and Takahashi, T., Effect of crystal structure and lattice parameters of iron up to 300 kbar, *J. Appl. Phys.*, *38*, 272-276, 1967.

Masters, G., Observational constraints on the chemical and thermal structure of the Earth's deep interior, *Geophys. J. R. Astron. Soc.*, *57*, 507-534, 1979.

Mollett, S., Thermal and magnetic constraints on the cooling of the Earth, *Geophys. J. R. Astron. Soc.*, *76*, 653-666, 1984.

Pfann, W.G., *Zone Melting*, p. 212, Wiley, New York, 1958.

Ringwood, A.E., Composition of the core and implications for origin of the Earth, *Geochem. J.*, *11*, 111-135, 1977.

Sclater, J.G., C. Janpart, and D. Galson, The heat flow through continents and oceans, *Rev. Geophys. Space Phys.*, *18*, 269-311, 1980.

Stacey, F.D., The cooling Earth: A reappraisal, *Phys. Earth Planet. Inter.*, *22*, 89-96, 1980.

Stevenson, D.J., Applications of liquid state physics to the Earth's core, *Phys. Earth Planet. Inter.*, *22*, 42-52, 1980.

Stevenson, D.J., Models of the Earth's core, *Science*, *214*, 611-619, 1981.

Strong, H.M., Tuft, R.E., and Hannemann, R.E., The iron fusion curve and the γ-ε-ℓ triple point, *Metall. Trans.*, *4*, 2657-2661, 1973.

Verhoogen, J., Heat balance in the Earth's core, *Geophys. J.*, *4*, 276-281, 1961.

Verhoogen, J., *Energetics of the Earth*, pp. 51-65, National Academy of Sciences, Washington, D.C., 1980.

Williams, Q., R. Jeanloz, J. Bass, B. Svendsen, and T.J. Ahrens, The melting curve of iron to 250 GPa: A constraint on the temperature at the Earth's center, *Science*, *236*, 181-183, 1987.

Young, D.A., and R. Grover, Theory of the iron equation of state and melting curve to very high pressure, in *Shock Waves in Condensed Matter-1983*, edited by J. Asay, R.A. Graham, and G.K. Straub, pp. 65-67, Elsevier, Amsterdam, 1984.

MOTIONS OF THE EARTH'S CORE AND MANTLE AND VARIATIONS OF THE EARTH'S MAGNETIC FIELD REVISITED

Raymond Hide

Geophysical Fluid Dynamics Laboratory, Meteorological Office
(Met 0 21), London Road, Bracknell, Berkshire, RG12 2SZ, U.K.

Introduction

The so-called geomagnetic "geocentric axial dipole" (G.A.D.) hypothesis - that when averaged over a few thousand years the geomagnetic field at the Earth's surface is close in form to that of an imaginary magnetic dipole situated at the geocentre and aligned along the Earth's rotation axis - has been used successfully in geological studies, including the testing in the 1950's of Wegener's proposals concerning continental drift by palaeomagnetic workers. But more attention may yet have to be given to devising palaeomagnetic and other tests of the limitations of the G.A.D. hypothesis. Changes in the geomagnetic field on timescales of core motions, including the tendency for some features of the non-dipole field to migrate westward at a fraction of a degree of longitude per year, undoubtedly smooths non-axisymmetric features of the field to some extent. Complete smoothing, however, is likely to be a much slower process, as became evident in the 1960's when I argued, largely on general theoretical grounds that departures from axial symmetry in the thermal and mechanical boundary conditions imposed on core motion by various processes including deep mantle convection would produce distortions in the main geomagnetic field that would average out very slowly indeed, on geological timescales of millions of years. Non-axisymmetric features of the core-mantle interface would also influence the amplitude and other properties of the geomagnetic secular variation and the frequency of polarity reversals, both of which were also amenable to investigation by geomagnetic workers.

The IUGG is about to set up a special "Study of the Earth's Deep Interior" (SEDI), recognizing that the time is probably ripe for a highly interdisciplinary concerted effort to be made to investigate in some detail the structure, dynamics and composition of the seventy-five percent of the whole Earth that lies beneath the upper mantle. The notion of intermittent convection throughout the mantle interacting strongly with the core could serve as a useful paradigm, to be exploited and modified in the light of new knowledge. When I submitted the abstract of this talk (see first paragraph above) to the organizers of the meeting, it was my intention to see what further palaeomagnetic evidence could be found bearing on the G.A.D. hypothesis, but I am inclined now to discuss other matters. From very helpful correspondence with a number of leading palaeomagnetic workers, I have become convinced that they are well aware of the need for further quantitative work on the limitations of the G.A.D. hypothesis and need no further prodding from me. Indeed, there have been several papers in IAGA symposia at this IUGG General Assembly that bear on the subject, including some presented a few evenings ago at a stimulating session on the magnetic properties of lake sediments. So I will deal with other questions that fall within the scope of this symposium, thereby effectively introducing some of the major problems and topics being discussed by those of us involved in the SEDI programme.

Dynamo Theory and Planetary Magnetic Fields

The G.A.D. hypothesis appeared to gain support when starting in 1955 it was discovered by radioastronomers and space scientists that the dipole axes of the planets Jupiter and Saturn nearly coincide with their rotation axes. So it came as a surprise last year when Dr.N.F.Ness and colleagues of the NASA Goddard Space Flight Center found that Uranus's magnetic dipole field is apparently inclined to the rotation axis by about 60° and displaced from the centre of the planet by about 0.3 times its radius. These are very much bigger than the corresponding values for the Earth, Jupiter and Saturn. The effective dipole moment was found to be 0.23 Gauss R_u^3 where R_u = 25,600 km, corresponding to minimum and maximum fields at the surface of Uranus of ~0.1 and ~1.1 Gauss respectively. (1 Gauss = 10^{-4}T.) The rotation period of the magnetic field, which should be close to that of the deep interior of the planet, was found to be 17.29 ± 0.10 hours.

The eccentricity of Uranus's magnetic field should be accepted by theoreticians as presenting a strong challenge. In a strategy towards an interpretation of the phenomenon I have presented elsewhere (see The Observatory 1986, 106, 144-146), the rôle played by rotation in the generation of magnetic fields by self-exciting magnetohydrodynamic (MHD) dynamo action is the central consideration. Such action occurs in well-stirred electrically-conducting fluids of sufficiently large dimensions. By Cowling's theorem and its recent extensions, no magnetic field with an axis of symmetry can be maintained against ohmic dissipation by fluid motions, so non-axisymmetric magnetic fields and patterns of fluid motions will be characteristic features of self-exciting dynamos. Non-axisymmetric flow patterns are readily shown to be characteristic features of motions of a rapidly-rotating fluid of low viscosity when any magnetic field present is weak, even when the boundary conditions are axisymmetric, and therein lies what is possibly the principal rôle played by rotation in promoting dynamo action.

It can be argued that the magnetic field thus produced builds up in strength until Lorentz forces are comparable in magnitude with Coriolis forces and are thereby able to annul or reduce considerably their inhibiting effect on advective transport processes (e.g. heat transfer) associated with the agencies responsible for driving the fluid motions. This "magnetostrophic" hypothesis gives reasonable values for the strengths of the Earth's poloidal and toroidal magnetic fields in their region of origin, the Earth's molten iron core, and ancillary arguments indicate why one might expect significant dynamo action within Jupiter, Saturn, Uranus and Neptune to take place in their moderately conducting non-metallic regions. The observed strength of the effective magnetic dipole of Uranus, nearly two orders of magnitude greater than that of the Earth, is not inconsistent with these considerations.

The inclination of the effective field, its displacement from the centre of the system, and the magnitude of higher order multipole components of the field will depend inter alia on the form of the departure from axial symmetry of the flow pattern. If the flow consists of many waves or eddies distributed more or less uniformly in azimuth, then the induced effective magnetic dipole field, seen at great distances from the surface of the planet, will at all times be approximately centred and aligned with the rotation axis, as in cases of Earth, Jupiter and Saturn. But if departures from axial symmetry in the flow are confined to a limited range of azimuth, then the magnetic dipole field will be highly eccentric in position and orientation. There is ample evidence from experiments carried out for other purposes in my laboratory, and also from studies of natural systems (for example, the Jovian Great Red Spot and other long-lived atmospheric eddies), that both types of flow are possible depending on circumstances, even when the boundary conditions are axisymmetric, and that the second type of flow is not highly improbable. Hence, if the second type of flow is occurring in the deep 'ocean' of water, ammonia and methane that we are told underlies the hydrogen, helium and methane atmosphere of Uranus, then a highly-eccentric magnetic field would be expected. (Another possibility - but in my opinion an improbable one - is that Uranus's magnetic field has been caught in the act of reversing its polarity. Yet another line of attack would be to consider whether Uranus's eccentric magnetic field is akin to those of typical stellar "oblique rotators".) It is possible but not likely that more than one planetary magnetic field is produced by flow of the second type. So Neptune's magnetic field might be expected to show approximate alignment with the rotation axis when Voyager 2 makes the measurement in a few years' time.

On the foregoing interpretation of the eccentricity of Uranus's magnetic field $\underset{\sim}{B}$ one would expect good alignment between $\underset{\sim}{B}$ and the rotation axis of the electrically-conducting fluid core when $\underset{\sim}{B}$ is averaged over a sufficiently long period of time, possibly centuries or more. But a method can be proposed for determining rather more quickly the direction of the rotation axis of the fluid core. Scale analysis of the equations of magnetohydrodynamics indicates that in the outer reaches of the core, where the toroidal part of $\underset{\sim}{B}$ should be comparatively weak, approximate geostrophic balance between the horizontal components of the pressure gradient and Coriolis forces should obtain. In these circumstances, the corresponding cross-equatorial flow would be negligible, implying no significant advection of lines of force of the poloidal magnetic field from one "geographic" hemisphere to the other. So the number of intersections of $\underset{\sim}{B}$ with each such hemisphere cannot change on timescales much less than that of free ohmic decay, and in principle it should be possible to determine the orientation of the geographic equator and therefore the direction of the rotation axis of the electrically-conducting fluid core of Uranus if and when measurements of secular changes in $\underset{\sim}{B}$ become available.

The Determination of Core-mantle Coupling from Geophysical Data

We now turn to another area of study of relevance in the investigation of the structure and dynamics of the Earth's deep interior. Geophysicists accept that the irregular "decade variations" in the length of the day of up to about 5×10^{-3}s are a manifestation of angular momentum exchange between the core and mantle. Concomitant fluctuating torques at the core-mantle interface are due to time-varying fluid motions in the liquid metallic core. The implied stresses at the core-mantle interface arise as a result of the action of (a) tangential viscous stresses in the Ekman-Hartmann boundary layer, (b) tangential Lorentz forces associated with the interaction of electric currents in the weakly-conducting lower mantle with the magnetic field there, and (c) normal pressure forces on bumps (i.e. departures in shape from axial symmetry) on the core-mantle boundary. The investigation of the relative effectiveness of these agencies is clearly a matter of geophysical interest.

The contribution of viscous stresses is unlikely to be significant except under extreme assumptions about the coefficient of viscosity of the core. For this reason, Bullard proposed in the 1950's that electromagnetic coupling must be the principal agency. Subsequent refinements in theoretical models of electromagnetic coupling have strengthened the original case for invoking that mechanism, but both qualitative and quantitative difficulties remain, the latter being associated with assumptions concerning the strength of the toroidal part of the geomagnetic field in the outer reaches of the core and the distribution of electrical conductivity in the lower mantle.

In the 1960's, in an attempt to overcome these difficulties and for other reasons, I proposed the idea of topographic coupling, arguing that the magnitude of the stresses implied by the amplitude and timescale of the decade variations in the length of the day might easily be accounted for if there are bumps on the core mantle boundary of height $\underline{h}$ no greater than about a kilometer and possibly less. Such bumps could easily be maintained by viscous stresses associated with deep convection in the mantle (not a popular idea in the 1960's and 1970's, when mantle convection was generally regarded as being confined to the top 700kms, but one which is now accepted by many geophysicists). How bumps and horizontal temperature variations at the core-mantle interface due to deep mantle convection influence core motions, thereby distorting the Earth's magnetic field, poses important questions in geophysical fluid dynamics, the further investigation of which should be of considerable theoretical and practical significance in the near future. But it is also of interest to consider whether direct estimates of topographic coupling can be obtained more or less directly from geophysical data.

A method for doing this has recently been proposed (Hide,R., Quart.J.Roy.Astron.Soc. 1986, 27, 3-20) and its practical applicability is now being studied by a group consisting of R.Clayton, B.Hager and M.A.Spieth of Cal.Tech., C.Voorhies of the Goddard Space Flight Center, and myself. From geomagnetic secular variation data, fields of horizontal motion just below the core-mantle interface are obtained on the basis of a method that exploits Alfvén's frozen magnetic flux theorem plus additional reasonable hypotheses concerning the dynamics of the flow. Horizontal pressure gradients are obtained from these hypothetical velocity fields on the basis of the geostrophic relationship, which should apply in the outer reaches of the core, where the largest ageostrophic contribution, from the Lorentz force is probably no more than about 10^{-2} times the Coriolis force in magnitude. Gravity and seismic data incorporated in various rheological models of the mantle provide hypothetical topographic maps of the core-mantle interface. The "decade" contribution to changes in the length of day and corresponding changes in the direction of the Earth's rotation axis (polar motion) are obtained from astronomical observations of the Earth's rotation when allowance has been made for tidal effects and short-term contributions due to the atmosphere.

First results are encouraging, for they show that for one epoch studied to date topographic coupling could account for the observed changes in the Earth's rate of rotation both qualitatively and quantitatively, without having to invoke extreme models of core-mantle interface topography and fields of core motions. Specifically, values of the effective topographic height h of up to about 0.5 km but no more are implied by these calculations of topographic coupling. This is in keeping not only with certain fluid-dynamical arguments but also with recent determinations from VLBI (Very Long Baseline Interferometry) data of the amplitude and phase of the free core nutation. But such values of h are less, by a factor of nearly ten, than topographic heights proposed in some seismological studies. Now, the effective topographic height h should be the same as the actual height if the metallic core is in direct contact with the lower mantle. However, as Professor Don Anderson has pointed out to me, some geophysicists have suggested that there might be a stable layer of poorly-conducting liquid "slag" separating the metallic core from the overlying mantle, in which case the effective topographic height h to be used in calculations of topographic coupling and the properties of the Earth's free nutation could be less than the actual height and possibly much less, though directly related to it. Efforts to resolve this matter will clearly be of great interest during the first phase of the SEDI programme.

TYPES OF LIQUID CORE MOTIONS COMPATIBLE WITH THE OBSERVED GEOMAGNETIC SECULAR VARIATIONS OF SEVERAL HUNDRED YEARS

Takesi Yukutake and Yukiko Yokoyama

Earthquake Research Institute, University of Tokyo
Bunkyo-ku, Tokyo, 113 Japan

Abstract. The geomagnetic field is decomposed into standing and drifting parts of which the drifting field has a special spatial structure consisting of two types of fields, field of sectorial harmonic type (m, m) and that of the harmonics (m+1, m). When the dipole field and the toroidal field of (n=2, m=0) are assumed as primary fields, only three kinds of interaction are capable of inducing the observed poloidal fields, first interaction of toroidal motions with the dipole field, second interaction of poloidal motions with the toroidal field and third that of poloidal motions with the dipole field. Induced modes for these three interaction processes are computed and relative amplitudes of the induced poloidal modes are estimated to compare with the observed modes.

Among possible processes to explain the observed standing and drifting field, a proposed model is that the standing field is produced by laminar flows in a boundary layer near the core surface through interaction with the dipole field, whereas the drifting field of sectorial harmonics is generated by convection type motions of (m+1, m) mode, which is antisymmmetric to the equator, interaction with both the toroidal and the dipole field in a deeper region of the core.

Introduction

The geomagnetic field is believed to be generated through electromagnetic induction processes in the earth's core by interaction of fluid motions with the magnetic field. Many efforts have been concentrated on examining the geomagnetic secular variations to infer the fluid motions in the core [see Benton, 1979; Muth and Benton, 1981; Yukutake, 1981; Gubbins, 1982; Madden et al., 1982; Whaler, 1982; Muth, 1983; LeMouel, 1984; LeMouel et al., 1985; Voorhis, 1986; Gire et al., 1986]. While the axisymmetric field like the axial dipole field is likely to be caused by combined processes of different types of induction, generation of a non-axisymmetric field is supposed to be more straightfoward through interaction of a single type of fluid motion with a specific kind of magnetic field. Subsequently the non-axisymmeric field is usually examined to infer the fluid velocity.

The secular variations cover a wide range of periods. Recently variations of a few tens of years are intensively investigated to infer the fluid motions at the shallow part of the core on the basis of frozen flux approximation [see, for example, Muth and Benton, 1981; Bloxham and Gubbins, 1986]. For this period range, the skin depth is so shallow as about 20 km for the currently accepted value of the electrical conductivity that only the field generated near the top surface of the core is considered to be observable. Furthermore the magnetic Reynolds number becomes so large that the magnetic diffusion can be ignored safely. In this case time variation in the magnetic field is caused by advection of magnetic lines of force. The geomagnetic secular variations as well as the main field are extrapolated down to the core surface and the fluid motions are estimated.

In this paper, however, we discuss about the fluid motions creating the secular variations of hundreds to thousand years. A characteristic feature of the geomagnetic variation of this time scale is the westward drift. The non-axisymmetric field is known to drift westwards with a velocity of 0.2 to 0.3° /year. This phenomenon is interpreted in two ways. Bullard et al. [1950] consider that the upper part of the core flows westwards as a whole relative to the mantle along with the magnetic fields that are frozen in the drifting core. The other interpretation is that the westward drift is manifestation of propagation of hydromagnetic waves in the core [see Hide, 1966; Braginskiy, 1967, 1980]. In this study we attempt to find what types of fluid motion are compatible with the observed fields that have been rotating over hundreds of years.

Although the westward drift is a well known

feature of the geomagnetic secular variation, examination of magnetic data since the 16th century indicates that the field consists of not only the drifting field but also the field standing at the same locality [Yukutake and Tachinaka, 1969]. From examination of time variation in the spherical harmonic coefficients, Gauss coefficients, of the field models, the drifting field has been found to be expressed by particular harmonics, whereas the standing field by all the harmonics [see Yukutake, 1987]. It is likely that the drifting field further consists of not only westward drifting but also eastward drifting fields. Extrapolation of the field observed at the surface since the 16th century down to the core mantle boundary leads Bloxham [1986] to conclude that most of the field is stationary and only limited types of field are drifting westwards.

At the surface of the earth the dipole field is overwhelmingly predominant in comparison with the fields expressed by other harmonics. This is the case even in the core if the region is limited near its surface. Therefore some of the observed features are possibly created through interaction of fluid motions near the surface with the dipole field. Both types of velocity, poloidal and toroidal velocity, are capable of inducing the observable poloidal types of magnetic fields through interaction with the dipole field [Yokoyama, 1987; Yokoyama and Yukutake, 1987]. Interaction with toroidal fields is another possible process to produce the poloidal field. In this case only the poloidal velocity is capable of inducing the poloidal field. There are still arguments whether intense toroidal fields exist or not. An intense toroidal field would create magnetic fields of observable magnitudes at the earth's surface. Examination of possible types of fluid motions could be a step towards clarifying the problem.

In this paper, we first describe characteristic features of the drifting and the standing fields, and discuss possible types of fluid motions appropriate to explain the observed features of the magnetic fields.

Drifting and Standing Fields

Figure 1 shows the vertical component of the non-dipole part of the geomagnetic field at the surface of the earth for two different epochs, 1770 and 1965 [see Yukutake, 1979]. The charts are synthesized from spherical harmonic models. This figure suggests that the geomagnetic field consists of two kinds of field, standing and drifting fields [Yukutake and Tachinaka, 1969]. Although the westward drift is a well known feature of the geomagnetic secular variation, it is seen that most of field patterns have stayed almost at the same place during the period from 1770 to 1965 except for a few foci such as a negative anomaly covering the western part of Africa in 1965 [Yukutake and Tachinaka, 1968, 1969]. The African negative focus was situated in the Indian Ocean in 1600. It moved westwards with a mean velocity of 0.28° /year to reach the west coast of Africa in 1965.

The drifting and the standing field can be separated by analysing the Gauss coefficients. In the drifting-and-standing field model, the Gauss coefficients g_n^m and h_n^m can be expressed by two terms,

$$g_n^m = F_n^m \cos(\alpha_n^m) + K_n^m \cos m v_n^m (t - \tau_n^m),$$

$$-h_n^m = F_n^m \sin(\alpha_n^m) + K_n^m \sin m v_n^m (t - \tau_n^m),$$

where F_n^m and α_n^m represent amplitude and phase angle of the standing field, whereas K_n^m, v_n^m and τ_n^m are amplitude, velocity and time phase of the drifting field respectively. We call this two mode model which consists of one standing and one drifting term. When g_n^m and h_n^m are plotted on a diagram with g_n^m in the abscissa and h_n^m in the ordinates as shown in Fig. 2, the standing field is expressed by a fixed vector drawn from the origin and the drifting field by a rotating vector. The synthetic field is expressed by a circle around the tip of the standing vector in Fig. 2. In this diagram westward drift is expressed by a clockwise rotation and estward drift by a counterclockwise rotation.

Drifting Harmonics

For the geomagnetic field since the 16th century, considerable number of spherical harmonic models are available [Barraclough, 1978]. The models for the earlier epochs are terminated at the maximum degree and order of 4. From examination of time variations of each harmonic coefficient, it is noted that the drifting field is contained only in particular harmonics, while the standing field in all harmonics [Yukutake, 1987]. Figure 3 shows an example for degree (n=2) and order (m=2), where the Gauss coefficients are averaged at 50 year intervals and plotted. Clockwise rotation is clear in this diagram, indicating that the field expressed by this harmonic term drifts westwards. Assuming the two mode model, the standing and the drifting fields have been determined. The double circle represents the end point of the standing field vector, and the large circle with its center at the double circle is the synthesized orbit from the best fitted model.

Although the general feature of the observed variation is approximated by the synthesized orbit, a large circle in the present case, there still remains considerable discrepancy. The observed orbit is more like an ellipse rather than a circle. The discrepancy has been greatly reduced by introducing another type of field into the model, field drifting eastwards [Yukutake, 1987].

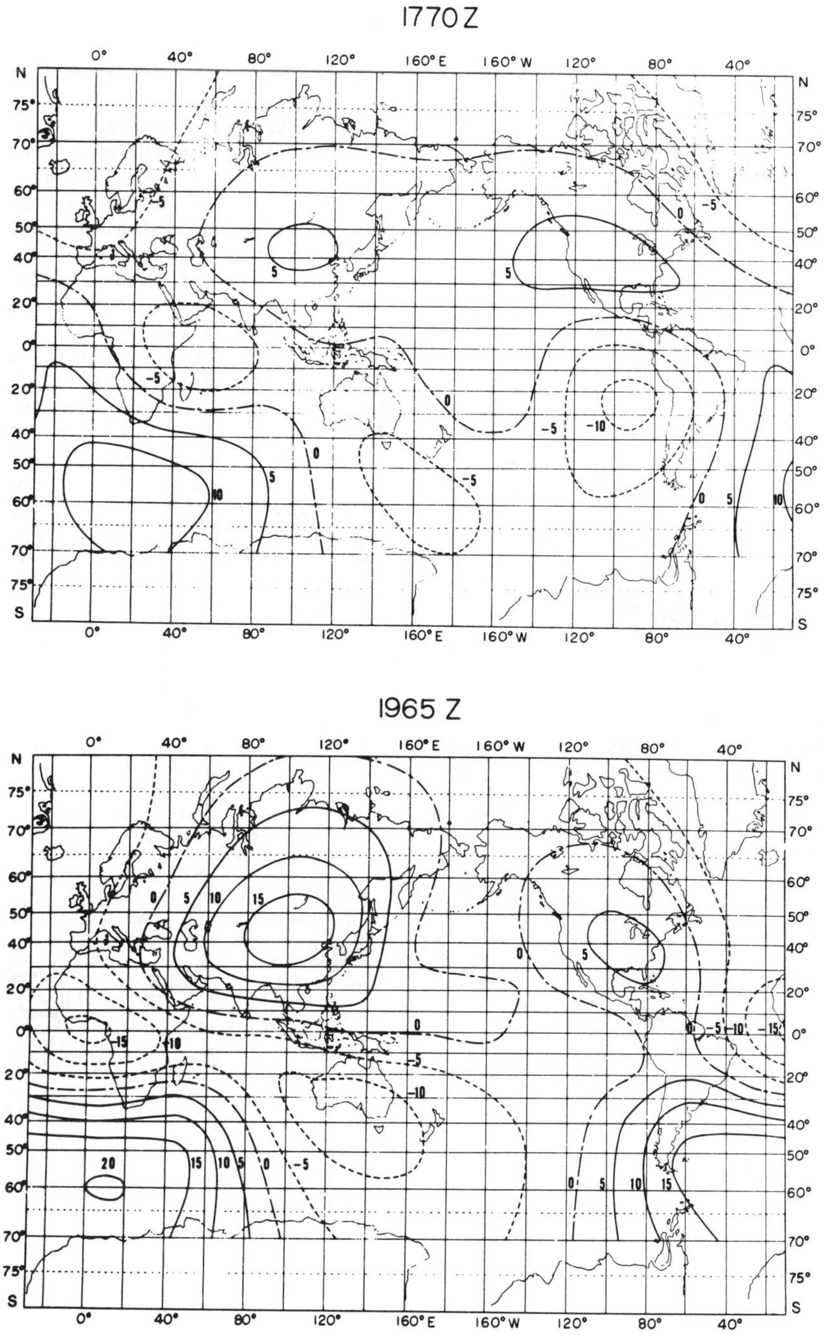

Fig. 1. The non-dipole vertical component of the geomagnetic field at the surface of the earth for two different epochs, 1770 and 1965. Contour intervals are 5000 nT.

However, the eastward drifting field is small in comparison with the westward drifting one except for the equatorial dipole component (n=1, m=1), only to modify the clockwise rotating circluar orbit into an ellipse. Accordingly in this paper we will concentrate our discussion on the induction processes for the standing and the westward drifting field.

The harmonic component n=m=2 shown in Fig. 3 is a typical example mainly consisting of the drift-

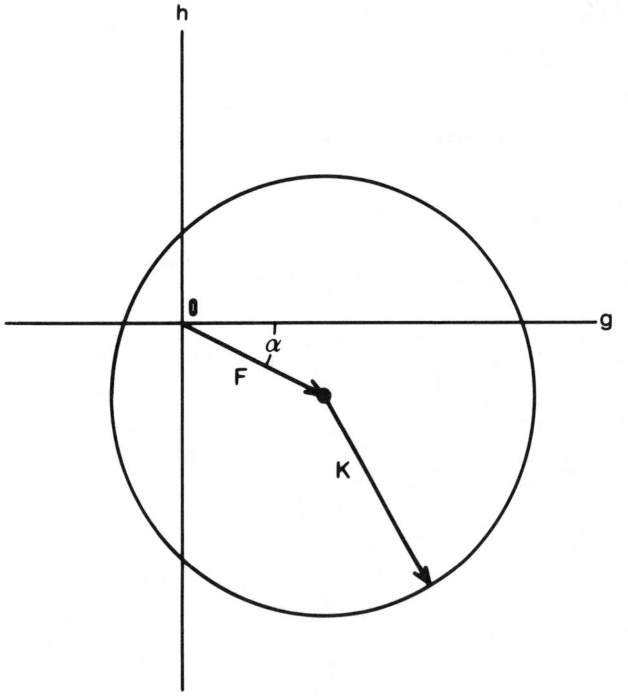

Fig. 2. Representation of two mode model, the standing and the drifting field, on a g-h diagram. A fixed vector drawn from the origin represents the standing field, while a vector depicting a circular orbit represents the drifting field. The westward drift is expressed by a clockwise rotation.

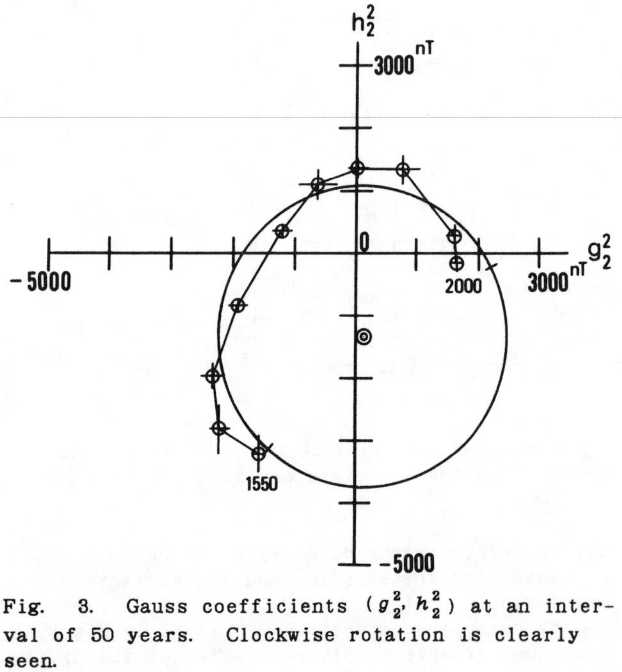

Fig. 3. Gauss coefficients (g_2^2, h_2^2) at an interval of 50 years. Clockwise rotation is clearly seen.

ing field. But not all the harmonics show such clear drifting nature. It is difficult to see such drifting feature in Fig. 4 which shows time variations in the harmonic n=4 m=2. From examination of the Gauss coefficients up to n=m=4, it is found that the drifting nature is seen only in particular harmonics. The sectorial terms, defined as those with the same degree and order, n=m, are the harmonics in which the drifting field is predominant. Except for the sectorial harmonics limited number of harmonics show drifting feature. Among other harmonics the component n=2 m=1 shows a remakable drifting nature. The amplitude of the drifting field is above all large in this component. Drifting feature is also recognized in the harmonic component n=4 m=3. From these it is likely that the drifting field is contained in the harmonics with n=m+1, m=m as well as in the sectorial harmonics n=m. In other harmonics, however, no clear rotational orbit is seen when the Gauss coefficients are plotted on the g-h diagram. This makes a clear contrast to the standing field that is contained in all harmonics.

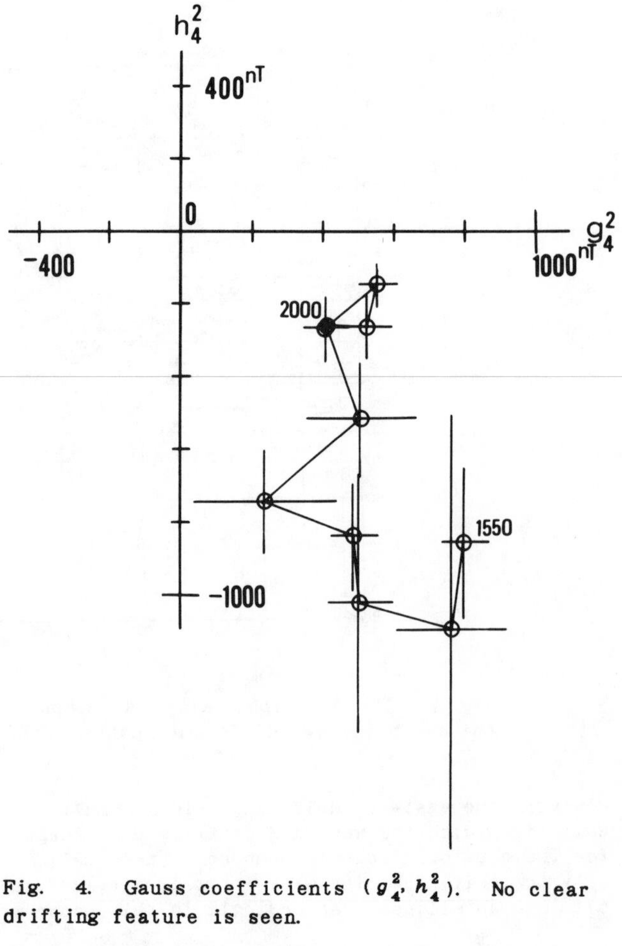

Fig. 4. Gauss coefficients (g_4^2, h_4^2). No clear drifting feature is seen.

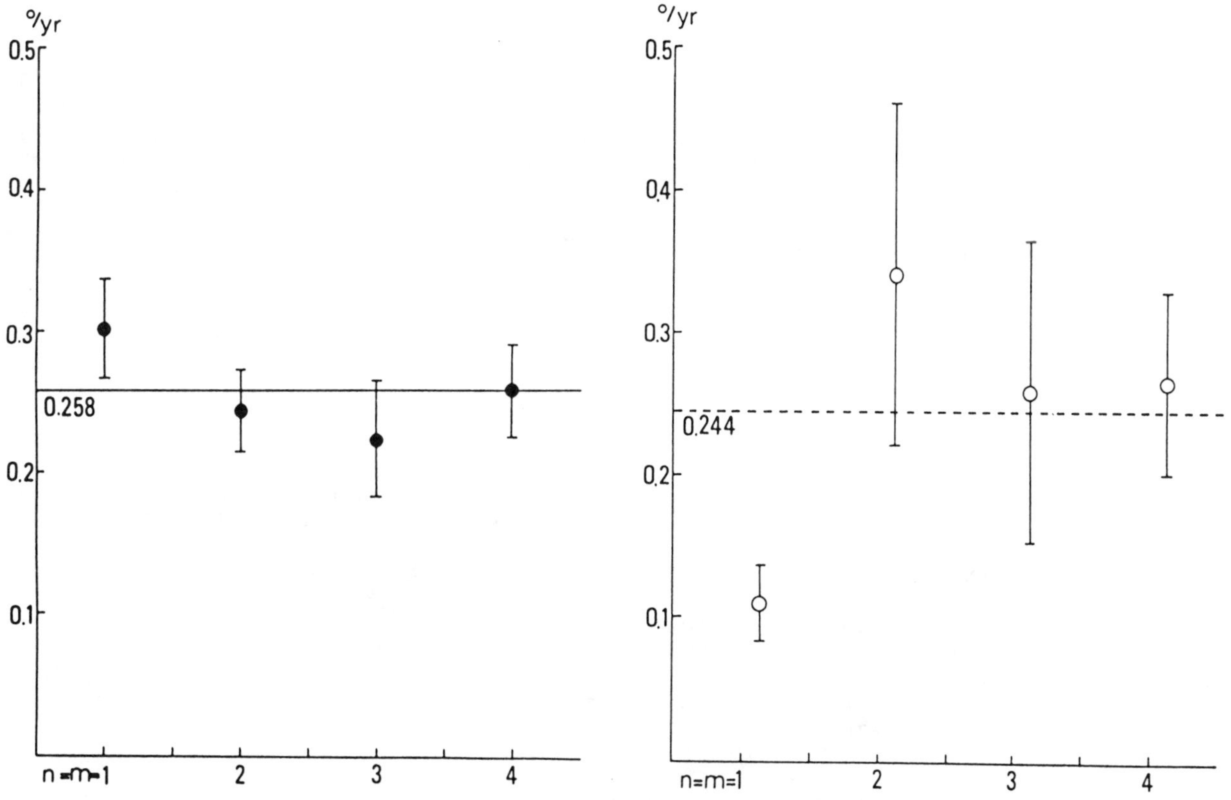

Fig. 5. Drift velocities of sectorial terms (m, m). Solid circles are drift velocities of the westward drifting component, whereas open circles those of the eastward drifting one. The mean velocity is 0.26° /year for the westward drift, and 0.24° /year for the eastward drift.

In other words, it can be said that the drifting field is made up from particular harmonics such as sectorial harmonics and those of n=m+1 m=m, whereas the standing field from all the harmonics. This implies that the drifting field has a special spatial structure.

It would be useful to mention the drift velocity of the drifting field briefly. The drift velocities of the sectorial terms are rather uniform. Their velocities of the westward drifting component, plotted in Fig. 5 by solid circles, range from 0.23 to 0.30°/year with a mean of 0.26°/year. Open circles are the velocities of the eastward drifting components. They show larger scatter, but the mean velocity becomes 0.24° /year, which is similar to that of the westward components.

Spatial Power Spectra of Standing and Drifting Fields

The spatial power spectrum is often used to estimate the depth of the source at which the observed field is generated [e.g. Lowes, 1974; McLeod and Coleman, 1980; Langel et al., 1982]. The power spectrum for the field observed at the surface of the earth usually decreases with increase of degree of harmonics. By downward-extrapolation, the power of the higher degree harmonics increases more rapidly than of the lower degree harmonics, and at a certain depth the spectrum becomes flat irrespective of the harmonic degree. The depth is assumed to give the depth of the source. Although only a limited number of harmonics are available in the present case, we have attempted to apply the technique to the drifting field and the standing field separately.

The power spectrum (R_n) for each harmonic degree n has been computed by the following equation presented by Lowes [1974],

$$R_n = (n+1)\left(\frac{R}{r}\right)^{2n+4}[(g_n^m)^2+(h_n^m)^2]$$

up to degree 4 for the standing field and for the westward drifting field respectively [Yokoyama, 1987], where R is the radius of the earth. The results are shown in Fig. 6 for the standing field, and in Fig. 7 for the westward drifting field. Circles are the power at the surface of the earth.

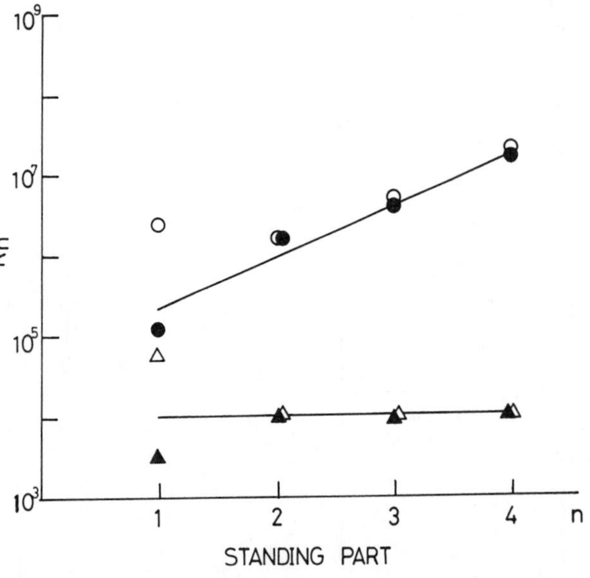

Fig. 6. Spatial power spectrum of the standing field. Triangles are the spectra at the surface of the earth and circles at the surface of the core. Open circles and open triangles are when zonal harmonics are included, while solid circles and solid triangles are when they are excluded. The spectra are almost flat at the surface of the earth whereas they increase with harmonics at the surface of the core.

while triangles represent the power at the surface of the core. Since separation of the drifting and the standing field is impossible with zonal harmonic terms m=0, special care is needed to treat these harmonics. Open circles and triangles represent power spectra when zonal harmonics are included in the calculation of power, whereas solid circles and triangles are those when they are not.

Figure 6 shows that the power spectrum for the standing field is already flat at the surface of the earth. However, this does not necessarily mean that the source of the standing field lies at the surface of the earth. The power obtained here is by far higher, about three order of magnitudes higher, than that of the crustal field estimated by Meyer et al. [1983] with realistic magnetization of the crust assumed for a geological structural model of global scale. Furthermore the standing field is likely to be subjected to fluctuations with time. When the field is extrapolated down to the core mantle boundary, the power increases with degree. This implies that short wavelength fields predominate at the surface of the core. If we extrapolate the field down into the deeper core, the tendency will be enhanced and the power diverges at higher degrees. At this moment, we tentatively assume that the standing field is produced at very shallow part of the core where the diverging tendency is the least.

On the other hand, the power of the westward drifting field plotted in Fig. 7 is likely to decrease with increase of degree at the surface of the earth. When extrapolated down to the core mantle boundary, the spectrum becomes flat. This may suggest that the source is shallow in the core.

From above, we assume in this study that the source of the standing field lies very near the surface of the core with higher power contained in higher harmonics. The drifting field, on the other hand, is produced also at shallow depths, but at somewhat deeper region than the standing field.

Induction Equation

The equation governing the process of generating the magnetic field is the induction equation.

$$\frac{\partial \mathbf{B}}{\partial t} + \frac{1}{\sigma\mu} \operatorname{curl}^2(\mathbf{B}) = \operatorname{curl}(\mathbf{v} \times \mathbf{B}). \qquad (1)$$

where $\mathbf{B}$ is the magnetic flux density, $\mathbf{v}$ the velocity of fluid, σ the electrical conductivity, and μ the magnetic permeability. Let $\mathbf{B}$ be

$$\mathbf{B} = \mathbf{B}_o + \mathbf{b},$$

where $\mathbf{B}_o$ is the primary field, and $\mathbf{b}$ the field generated through interaction of small velocity $\mathbf{v}$

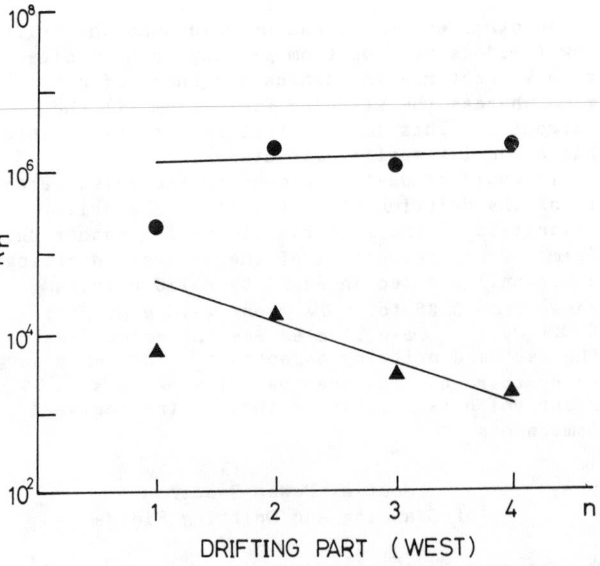

Fig. 7. Spatial power spectrum of the westward drifting components. The spectra become almost flat at the surface of the core.

with the field B_0. Ignoring product terms of small quantities, we have

$$\frac{\partial b}{\partial t}+\frac{1}{\sigma\mu}\mathrm{curl}^2(b)=\mathrm{curl}(v\times B_o). \qquad (2)$$

If the standing field does not change with time, equation

$$\frac{1}{\sigma\mu}\mathrm{curl}^2(b)=\mathrm{curl}(v\times B_o) \qquad (3)$$

should be solved to describe the induction process for the standing field. For the drifting field, on the other hand, we may take the time varying part as

$$v=v_o exp[im\omega t], \quad b=b_o exp[im\omega t].$$

Then the equation becomes

$$im\omega b_o+\frac{1}{\sigma\mu}\mathrm{curl}^2(b_o)=\mathrm{curl}(v_o\times B_o). \qquad (4)$$

If we take the angular velocity of the westward drift for ω, and the radius of the core for the length scale a, the induction parameter $(ka)^2[=\sigma\mu\, im\omega a]$ becomes of the order of 10^2. Similarly the magnetic Reynolds number $R_m[=\sigma\mu a v_o]$ becomes of the order of 10^2 for the observed drift velocity. This indicates that the diffusion term plays negligible part in the case of drifting field.

Whichever case may be considered, standing field or drifting field, computation of the induction term, the righthand side of eq.(1), is of primary importance. Yokoyama and Yukutake [1987] computed modes of induced fields through the induction term in general cases. Among them here we discuss interaction of fluid velocity with two specific types of fields, the dipole field and the toroidal field.

The dipole field dominates over other harmonics at the earth's surface. In the deep interiors, relative importance of the higher harmonics increaces and the diople field becomes less pronounced. Nevertheless the predominance of the dipole field still continues down to the core mantle boundary. Figure 8 shows the maximum intensity of the vertical component derived from zonal harmonics following the equation,

$$Z_{max}=(n+1)\left(\frac{R}{r}\right)^{n+2}g_n^0,$$

where R is the radius of the earth. Solid circles denote the intensity at the surface of the earth, and open circles the intensity at the core mantle boundary. At the earth's surface the vertical in-

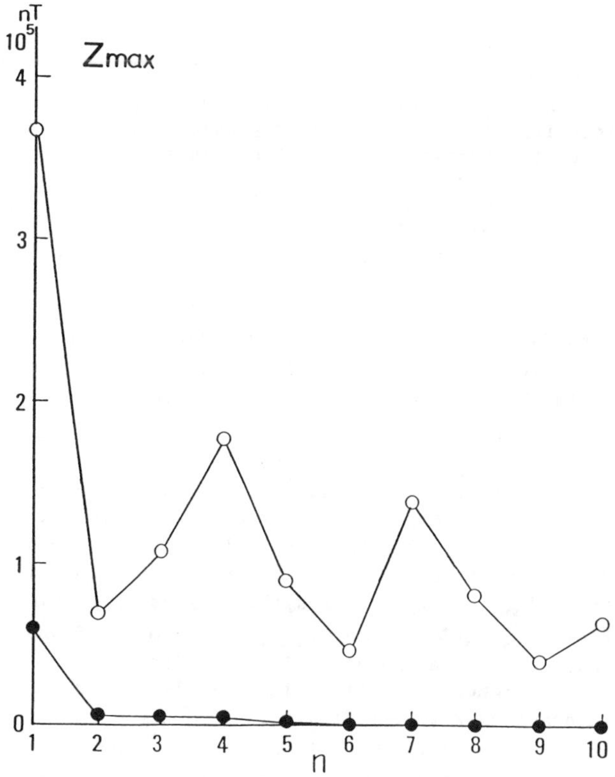

Fig. 8. The maximum magnitude of the vertical component produced by zonal harmonics. Solid circles are the magnitudes at the surface of the earth, while open circles at the core surface. The dipole is predominant not only at the surface of the earth but also at the surface of the core.

tensity of the quadrupole field that is the largest among the higher harmonics is about 10 percent of the dipole field. At the core mantle boundary higher harmonic field is smaller than 50 percent of the dipole field. Even at the depth of 350 km from the core surface the dipole field is predominant, the largest higher harmonic field being about 66 percent of the dipole field. Accordingly in the shallowest part of the core, at least down to a few hundred kilometers depth, the dipole field plays the most important role as the inducing field.

Another candidate to be considered for the inducing field is a toroidal field that is supposed to be produced through interaction of a rotational shear motion with the dipole field. If the toroidal field is as intense as suggested by the strong field model, the observed magnitude of poloidal field can be easily induced by interaction of convection type motions in the deeper part of the core. Here we discuss the induction by the toroidal field of n=2 m=0 with intensity of a few hundred gauss.

Modes of Induced Field from the Dipole and the Toroidal Fields

Since interaction of the toroidal type of velocity with the toroidal magnetic field only induces the toroidal field, three kinds of interaction are possible to produce the *observable poloidal types of magnetic field* from the dipole and the toroidal field of n=2 m=0.

Let v be the velocity, and B, b be the inducing and the induced magnetic field repectively. Subscripts $S_{n,m}$ and $T_{n,m}$ denote poloidal and toroidal types of degree n and order m. Then interaction of toroidal velocity $v_{Tn,m}$ with dipole field $B_{S1,0}$ induces three types of field, one poloidal field $b_{Sn,m}$ and two toroidal fields $b_{Tn-1,m}$ and $b_{Tn+1,m}$. It is noted that the induced poloidal field has the same degree and order as the inducing velocity. Interaction of poloidal velocity $v_{Sn,m}$ with the toroidal field $B_{T2,0}$ produces five types of field, two types of poloidal field $b_{Sn-1,m}$, $b_{Sn+1,m}$, and three types pf toroidal field $b_{Tn-2,m}$, $b_{Tn,m}$ and $b_{Tn+2,m}$. The mode of the induced poloidal magnetic field are (n-1, m) and (n+1, m), whereas the inducing velocity is (n, m). Interacting with the dipole field $B_{S1,0}$, the poloidal velocity $v_{Sn,m}$ induces $b_{Sn-1,m}$ and $b_{Sn+1,m}$, the same types of poloidal field as induced by interacting with the toroidal field. This process also generates a toroidal field of $b_{Tn,m}$. These three kinds of interaction are summarised in Table 1.

Since the standing field consists of all harmonics, any of the above three types of interaction can be the candidate to explain the observed standing field. However, generation process of the drifting field may be restricted since it consists of limited number of particular harmonics. In order to produce the sectorial mode of the drifting field n=m, two kinds of velocity are conceivable. One is the toroidal velocity $v_{Tm,m}$ that interacts with the dipole field $B_{S1,0}$. The other is the poloidal velocity $v_{Sm+1,m}$ that interacts with either of toroidal field $B_{T2,0}$ or the dipole field $B_{S1,0}$.

As can be seen in Table 1, the toroidal velocity $v_{Tm,m}$ produces only single mode of poloidal field $b_{Sm,m}$ through interaction with the dipole field. On the other hand, the poloidal velocity $v_{Sm+1,m}$ generates two modes of poloidal field, not only the sectorial mode $b_{Sm,m}$ but also the field $b_{Sm+2,m}$ by interacting with either the toroidal field $B_{T2,0}$ or the dipole field $B_{S1,0}$. Take the poloidal velocity $v_{S3,2}$ for example, it induces a typical drifting mode n=m=2, but at the same time it produces the poloidal field of $b_{S4,2}$. This implies that, if the drifting field in the sectorial term n=m=2 is produced by the interaction of the poloidal velocity $v_{S3,2}$ with the toroidal field $B_{T2,0}$, the poloidal field of $b_{S4,2}$ should be simultaneously induced as a drifting field. The observation, however, indicates that the drifting field is contained in the sectorial term n=m=2, but no significant drifting field in the harmonic term n=4 m=2. Generally speaking, simultaneous generation of two modes $b_{Sm,m}$ and $b_{Sm+2,m}$ is theoretically expected through interaction of the poloidal velocity $v_{Sm+1,m}$ with the toroidal field or with the dipole field, but the drifting field is observed only in the sectorial terms n=m, but not in terms n=m+2 and m=m. A quetion arises here whether we should exclude the poloidal velocity $v_{Sm+1,m}$ out of the candidates to produce the drifting field or not. There is a possibility, however, that the higher harmonics $b_{Sm+2,m}$ are too small to be observed, though induced. The question will not be solved until the amplitudes of the induced modes are calculated on an actual earth model.

TABLE 1. Three Types of Interaction to Induce Observable Poliodal Fields From the Dipole and The Toroidal Fields

	curl $(v \times B)$	
1)	$V_{Tn,m} \times B_{S1,0}$	→ $b_{Sn,m}$
		→ $b_{Tn-1,m}$, $b_{Tn+1,m}$
2)	$V_{Sn,m} \times B_{T2,0}$	→ $b_{Sn-1,m}$, $b_{Sn+1,m}$
		→ $b_{Tn-2,m}$, $b_{Tn,m}$, $b_{Tn+2,m}$
3)	$V_{Sn,m} \times B_{S1,0}$	→ $b_{Sn-1,m}$, $b_{Sn+1,m}$
		→ $b_{Tn,m}$

Amplitudes of the Induced Modes

We have made a preliminary investigation of the induction process on a simplified earth model [Yokoyama, 1987; Yokoyama and Yukutake, in preparation]. For the interaction process of toroidal motions with the dipole field, we assume that the toroidal motions are confined within a surface layer of finite thickness. If the thickness of the layer is thin, solving eq. (3) for the steady model, we have the complex Gauss coefficient $b_n^m (= g_n^m + i h_n^m)$ approximately as

$$b_n^m \fallingdotseq -\frac{2im\delta}{n(n+1)(2n+1)} R_m V_{Tn,m} B_{S1,0}$$

where $V_{Tn,m}$ is the magnitude of the toroidal velocity and $B_{S1,0}$ the intensity of the dipole field. For the time varying model, solving eq. (4), we have approximately

$$b_n^m \fallingdotseq \frac{2im}{n(n+1)} \frac{R_m}{(ka)^2} \frac{\delta}{1-\delta}[1-O(1/ka)]V_{Tn,m}B_{S1,0}.$$

This indicates that the magnetic field produced by toroidal motions varying with time scale of the observed westward drift is about $(2n+1)/(ka)^2$ times smaller than the field induced in the steady model. Similar results are obtained for the field produced by interaction of poloidal motions with the toroidal field. In this case, too, the field induced by the poloidal velocity rotating with the observed drift velocity is $n(2n+1)/(ka)^2$ smaller than that in the steady case.

On the other hand, in the case of interaction by a poloidal motion, two modes of field are simultaneously induced, whether it may interact either with the toroidal field or with the dipole field. The poloidal velocity $v_{Sn,m}$ induces $b_{Sn+1,m}$ and $b_{Sn-1,m}$. When it interacts with the toroidal field, relative magnitude of the complex Gauss coefficients becomes approximately,

$$\frac{b_{n+1}^m}{b_{n-1}^m} \fallingdotseq \frac{n(n-1)(n-m+1)}{(n+1)(n+2)(n+m)}$$

which is about 0.1 for n and m that are smaller than 4. This indicates that the induced higher mode is one order of magnitude smaller than the lower mode. Similar results are obtained for the interaction with the dipole field.

In the case of the steady model, numerical calculation has been conducted for a more specific model in which appropriate forms of radial functions are assumed with the toroidal field and the poloidal velocity [Yokoyama, 1987]. The velocity is assumed, for instance, to take the maximum at r= 0.9 a, where a is the radius of the core. Figure 9 shows an example of the results. The amplitudes of the induced field through interaction of poloidal velocity $v_{Sn,m}$ with the toroidal field $B_{T2,0}$ are plotted with the degree (n) and order (m) of the velocity in the abscissa. The ordinates are the magnitudes of the induced field plotted in terms of $b_n^m/[B_{T2,0}V_{Sn,m}\sigma\mu]$, where b_n^m is the magnitude of the induced field, $B_{T2,0}$ the intensity of the toroidal field, and $V_{Sn,m}$ the maximum amplitude of the velocity. Numbers (n, m) beside the circles denote degree and order of the induced field. The figure illustrates that, although two modes of field, (1, 1) and (3, 1) for example, are induced by velocity of n=2 m=1, the higher mode (3, 1) is almost two order of magnitudes smaller than the lower mode (1, 1). Similarly the mode (4, 2) is about one order of magnitude smaller than the mode (2, 2) which are both induced by the velocity of

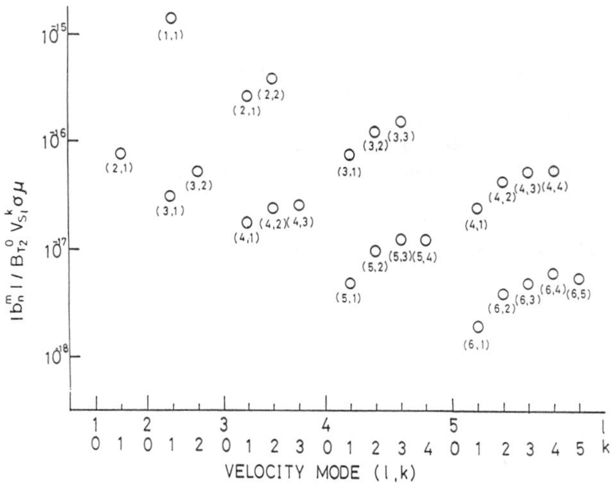

Fig. 9. Amplitude of induced fields through interaction of poloidal velocity $v_{Sn,m}$ with the toroidal field $B_{T2,0}$. The amplitude is plotted in terms of $B_{T2,0}V_{Sn,m}\sigma\mu$.

n=3 m=2. The figure suggests that sectorial terms n=m would predominate in the induced field, if the magnitude of the velocity were constant independent of its degree and order.

The above calculation indicates that the higher mode $b_{Sm+2,m}$ is much smaller than the lower mode $b_{Sm,m}$, which are both induced by interaction of the velocity $v_{Sm+1,m}$ with the toroidal field $B_{T2,0}$ or with the dipole field $B_{S1,0}$. This implies that the induction process described in this section can still be a candidate for explaining the drifting field that consists of sectorial harmonics, provided the induced higher mode is too small to observe, being below the noise level.

Discussions

The process of generating the standing and the drifting field is not yet well understood. The so called standing field could be simply another drifting mode with too small a velocity to be resolved during the observed time interval of several hundred years. In view of this, the standing field was once interpreted as a field residing in the deeper part of the core that was supposed to be rotating slightly faster than the mantle, while the drifting field was generated in the upper part of the core rotating westwards as a whole [Yukutake, 1981]. Recent analyses suggest that the eastward drifting component is likely to exist besides the westward drifting field and that the standing field is the field staying at the same place and varying with time scale of hundred years. If such relatively short time scale variations are real, it seems unnatural to ascribe the source of the standing field to the deeper part of the core. The generation process must be different from that of the drifting field.

It is often assumed that the depth at which the spatial power spectrum becomes flat gives a depth of origin of the field. Application of this technique suggests that the source of the standing field is shallower than the drifting field. The spectrum of the standing field becomes already flat at the surface of the earth and it increases with the harmonic degree at the core mantle boundary. From its large amplitude as well as its time variability, the standing field cannot be of crustal origin. In order to keep the diverging trend of the spectrum minmum, the source should be considered to reside at the very top of the core. Even in this case, higher harmonics predominate at the source level.

On the other hand, the spectrum of the westward drifting field converges at the surface of the earth, and becomes almost flat at the core mantle boundary. This also suggests that the source should be shallow. Since the core is electrically conducting that the drifting field would be electromagnetically shielded if it is generated in a deep interior of the core. The skin depth for the drift velocity 0.3° /year is estimated to be 180 km. This would give some concept about the depth of the source region for the drifting field. In this paper, we have assumed that it is shallow but somewhat deeper than the standing field.

Two possible sources are conceivable as primary fields from which the standing and the drifting field are induced, the dipole field and the toroidal field. At the core mantle boundary relative importance of higher harmonic field increases. Nevertheless the dipole field is still predominant near the surface of the core. The dipole field no doubt plays most important role in the induction process at the shallowest part of the core. On the other hand, there is still arguments about existence of the intense toroidal field. If it exists, the role of the toroidal field cannot be ignored in discussing the generation process of the field in the deeper part.

At the present stage, we are inclined to consider that the standing field is induced by interaction of toroidal motions in a boundary layer near the core surface with the dipole field. Several kinds of boundary layers are now considered. The Ekman layer is the layer which should certainly exists. In spite of large uncertainty of the viscosity of the core, the thickness of the layer is estimated to be less than 10 km. It is doubtful that such a thin layer could participate significantly in generation process of the standing field. Braginskiy [1984] proposes a magnetic diffusive layer in which variety of hydromagnetic phenomena would take place. The thickness of the layer is estimated to be about 35 km based on the skin depth estimate for the 65 year variations that are widely observed in the geomagnetic secular variations. Another type of the boundary layer is a stably stratified layer near the surface. Lighter materials which are contained as minor elements in the iron core are supposed to float up to the surface and form a stably stratified layer [see Braginskiy, 1984; Loper and Roberts, 1983]. Its thickness is estimated to be 70 km by Loper and Roberts, and 20 km by Braginskiy. Gubbins et al. [1982] suggest a possibility of forming a thermally stratified layer near the core mantle boundary in the case where the core mantle boundary cools more slowly than the core interior. The thickness is not well estimatable, ranging between 100 km and 1000 km. At the present time it is not yet decisive whichever case is most plausible, but a stratified layer is very likely to exist near the surface of the core. Laminar flows in the layer could produce the observable magnetic field by interacting with the dipole field. Layered motions of 5×10^{-3} cm/sec are capable of producing the observed magnitude of the standing field if the thickness of the layer is 35 km.

In the deep interior of the core, convective motions are supposed to occur. If an intense toroidal field of n=2 and m=0 type exists, convectional motions can induce poloidal fields observable at the earth's surface. Even if the intense toroidal field does not exist, interacting with the dipole field, the convectional motions induce the same type of poloidal fields as in the case of interaction with the toroidal field.

On the other hand, the drifting field consists mostly of sectorial terms (n=m, m), and possibly of harmonics (n=m+1, m). The drifting field can be produced by any of the above mentioned three types of interaction, interaction of toroidal motions with the dipole field, that of poloidal motions with the toroidal field or that of poloidal motions with the dipole field. Suppose the interaction of the toroidal motions with the dipole field. The drifting field of sectorial harmonics is induced by toroidal velocities of sectorial harmonics. This requires horizontal flows across the equator northwards at some places and southwards at other palces. Hydrodynamic consideration is necessary to justify this type of fluid motion, but this seems rather implausible in comparison with such a horizontal motion as is symmetric with respect to the equator. The drifting field of harmonics n=m+1, m=m is induced by the toroidal motion of the same harmonic degree and order. This motion is symmetric to the equator, which is not entirely unrealistic. Interaction of the poloidal velocity with the toroidal field explains the generation of the drifting field of the sectorial harmonics in a more natural way. Poloidal velocities of harmonics n=m+1, m=m induce the fields of sectorial harmonics. On the other hand, velocities of sectorial harmonics (n=m, m=m) or the harmonics (n=m+2, m=m) produce the field of n=m+1, m=m type. The velocities of the sectorial harmonics represent motions like cylindrical convections in a rotating system studied by Roberts [1968] and Busse [1970]. Interaction of the poloidal motions (n=m+1, m=m) with the toroidal field induces not only the poloidal field of sectorial harmonics (n=m, m=m) but also the field of (n=m+2, m=m). However, as discussed in a previous section, the induced harmonics (n=m+2, m=m) are much smaller than the sectorial harmonics (n=m, m=m). Therefore it is not unreasonable that the drifitng field is hardly recognized in the higher harmonics (n=m+2, m=m). Similar situation exists with the drifting field of (n=m+1, m=m) when induced by the motion (n=m+2, m=m). A field of (n=m+3, m=m) is induced simultaneously but much smaller.

At the same time, interacting with the dipole field, the poloidal velocities induce the same type of poloidal fields as in the case of interaction with the toroidal field. The situation is very similar to the toroidal case. Two modes are induced, but the higher mode is smaller. However, this has an important implication that, even if there is no intense toroidal field, such poloidal motions as discussed here can produce the observed drifting field provided the upper part of the core rotates westwards with the poloidal motions superposed.

From the above discussion alone, it is not determined whether the observed drifing field is produced through interaction of the convectional flows with the dipole field or with the toroidal field, or both. An advantage of the toroidal field model is, that it provides a slow hydromagnetic wave , a magnetostrophic wave, modified by the earth's rotation, which propagates along the toroidal field with a velocity of right order of magnitude to explain the observed drift velocity of the magnetic field, provided the intensity of the toroidal field is as intense as 100 gauss. Even though the drifting field is produced by this mechanism, the wave interacts with the dipole field and makes significant contribution to form the observed drifting field.

Concluding Remarks

At the present stage the investigation of the core motions relevant to the observed secular variations is not conclusive yet. Among the possible processes of producing the observed standing and drifting fields, our preferred interpretation is that the standing field is generated through interaction of laminar flows in the boundary layer near the core surface with the dipole field, whereas the major part of drifting field is produced by interaction of convection type motions in a deeper region with both the toroidal and the dipole fields.

Since the spatial structure of the standing field at the core mantle boundary is characterized by predominance of higher harmonics in the power spectrum, it is likely that the laminar flows are also dominated by motions of laterally short length scale. On the other hand, the convectional flows that induce the drifting field of the sectorial harmonic type are characterized by the velocity of $v_{Sm+1,m}$ type which is antisymmetric with respect to the equator. Another type of drifting field, $B_{Sm+1,m}$, is also possible to be caused through interaction with the toroidal field. In this case the velocity would take the form of $v_{Sm,m}$ which is independent of latitudes or of the coordinates in the direction of the rotational axis. However, a different interpretation is equally possible with this type of drifing field. Laminar motions of $v_{Sm+1,m}$ in a stratified layer could produce the field concerned by interacting with the dipole field.

Geomagnetic secular variations provide certain constraint on the possible types of fluid velocity in the core. In order to determine which type is the most likely, detailed examinations of hydrodynamics are needed.

References

Barraclough, D. R., Spherical harmonic models of the geomagnetic field, Geomagn. Bull. Inst. Geol. Sci. Edinburgh., 8, 1-66, 1978.

Benton, E. R., On fluid circulation around null-flux curves at earth's core-mantle boundary, Geophys. Astrophys. Fluid Dyn., 11, 323-327, 1979.

Bloxham, J., Models of the magnetic field at the core-mantle boundary for 1715, 1777, and 1842, J. Geohys. Res., 91, 13954-13966, 1986.

Bloxham, J. and D. Gubbins, Geomagnetic field analysis-IV. Testing the frozen-flux hypothesis, Geophys. J. Roy. astr. Soc., 84, 139-152, 1986.

Braginskiy, S. I., Magnetic waves in the earth's core, Geomag. Aeron., 7, 851-859E., 1967.

Braginskiy, S. I., Magnetic waves in the core of the Earth II, Geophys. Astrophys. Fluid Dyn., 14, 189-208, 1980.

Braginskiy, S. I., Short-period geomagnetic secular variation, Geophys. Astrophys. Fluid Dynamics, 30, 1-78, 1984.

Bullard, E. C., C. Freedman, H. Gellman and J. Nixon, The westward drift of the earth's magnetic field, Phil. Trans. Roy. Soc. London, A243, 67-92, 1950.

Busse, F. H., Thermal instabilities in rapidly rotating systems, J. Fluid Mech., 44, 441-460, 1970.

Gire, C, J. L. LeMouel and T. Madden, Motions at the core surface derived from SV data, Geophys. J. Roy. astr. Soc., 84, 1-29, 1986.

Gubbins, D., Finding core motions from magnetic observations, Phil. Trans. R. Soc. Lond., A, 306, 247-254, 1982.

Gubbins, D., C. J. Thomson and K. A. Whaler., Stable regions in the earth's liquid core., Geophys. J. Roy. astr. Soc., 68, 241-251, 1982.

Hide, R. Free hydromagnetic oscillations of the Earth's core and the theory of the geomagnetic secular variation, Phil. Trans. Roy. Soc. London, A., 259, 615-650, 1966.

Langel, R. A. and R. H. Estes, A geomagnetic field spectrum, Geophys. Res. Letters, 9, 250-253, 1982.

LeMouel, J. L., Outer-core geostrophic flow and secular variation of the Earth's geomagnetic field, Nature, 311, 734-735, 1984.

LeMouel, J. L., C. Gire and T. Madden, Motions at core surface in the geostrophic approximation, Phys. Earth Planet. Interiors, 39, 270- 287, 1985

Loper, D. and P. H. Roberts, Compositional convection and the gravitationally powered dynamo, in Stellar and Planetary Magetism, ed. A. M. Soward, Gordon Breach Sci. Publ (New York), 297-327, 1983.

Lowes, F. J., Spatial power spectrum of the main geomagnetic field, and extrapolation to the core, Geophys. J. R. astr. Soc., 36, 717-730, 1974.

Madden, T. and J. L. LeMouel, The recent secular variation and the motions at the core surface, Phil. Trans. R. Soc. London, A., 306, 271-280, 1982.

McLeod, M. G. and P. J. Coleman, Jr., Spatial power spectra of the crustal geomagnetic field and core geomagnetic field, Phys. Earth Planet. Int., 23, 5-19, 1980.

Meyer, J., J.-H. Hufen, M. Siebert and A. Hahn, Investigations of the internal geomagnetic field by means of a global model of the Earth's crust, J. Geophysics, 52, 71-84, 1983.

Muth, L. A., Can core surface velocities be determined from geomagnetic field models ? Developments towards a comprehensive theory, in Stellar and Planetary Magnetism, ed. A. M. Soward, Gordon Breach Sci. Pub (New York), 273-288, 1983.

Muth, L. A. and E. R. Benton, On the frozen flux velocity field at the surface of earth's core necessary to account for the poloidal main magnetic field and its secular variation, Phys. Earth Planet. Interiors, 24, 245-252, 1981.

Roberts, P. H., On the thermal instability of a rotating-fluid sphere containing heat sources, Phil. Trans. Roy. Soc. London, A., 263, 93-117, 1968.

Voorhis, C. V., Steady flows at the top of earth's core derived from geomagnetic field models, J. Geophys. Res., 91, 1244-12466, 1986.

Whaler, K. A., Geomagnetic secular variation and fluid motion at the core surface, Phil. Trans. R. Soc. Lond. A, 306, 235-246, 1982.

Yokoyama, Y., A mechanism of generating the geomagnetic secular variations, M. Sc. Thesis, Univ. Tokyo, (in Japanese), 1987.

Yokoyama, Y. and T. Yukutake, Calculation of induced modes of magnetic field in the geodynamo problem, submitted to J. Geomag. Geoelectr., 1987.

Yukutake, T., Review of the geomagnetic secular varitions on the historical time scale, Phys. Earth Planet. Interiors, 20, 83-95, 1979.

Yukutake, T., A stratified core motion inferred from geomagnetic secular variations, Phys. Earth Planet. Interiors, 24, 253-258, 1981.

Yukutake, T., A preliminary study on variations in the Gauss coefficients of the geomagnetic potential over several hundred years, Phys. Earth Planet. Interiors, 39, 217-227, 1985.

Yukutake, T., On the drifting and standing fields in the geomagnetic field, Rep. Heinrich-Herz Int., 21, 7-20, 1987.

Yukutake, T. and H. Tachinaka, The non-dipole part of the earth's magnetic field, Bull. Earthq. Res. Inst., 46, 1027-1074, 1968.

Yukutake, T. and H. Tachinaka, Separation of the earth's magnetic field into the drifting and the standing parts, Bull. Earthq. Res. Inst., 47, 65-97, 1969.

VARIATIONS OF THE VERTICAL BETWEEN MIZUSAWA AND WASHINGTON, D.C. AND THOSE OF THE GEOMAGNETIC FIELD

Chuichi Kakuta

International Latitude Observatory of Mizusawa, Mizusawa, Iwate, Japan

Abstract. Variations of the optical observations of time and latitude residuals are derived from two stations, the ILOM and the USNO. The latitude observation residuals show a correlation with local variations of the horizontal component and the *X* component of the geomagnetic field. However, time observation residuals show a correlation with local variations of the first time derivative of the geomagnetic field *Y* component.

Fluid motions in the fluid core are examined by taking into account the temperature gradient in the meridional direction near the CMB. The result shows the possibility of a Rossby type wave to induce variations of the disturbing potential and those of the vertical.

Introduction

Optical observations of astronomical time and latitude are based on the local vertical at the observation site. Local residuals of astronomical observations are derived by subtracting the common variations of the Earth Rotation Parameters (ERP), pole motion and UT1-TAI. The local residuals consist of errors of the observed star place and the ERP, effects of the observed environments such as atmospheric refraction and variations of the vertical. The International Latitude Observatory of Mizusawa (ILOM) and the United States Naval Observatory (USNO), Washington, D. C. are located nearly on the same latitude. Since the ILOM started the PZT observation in 1956, both observatories have continued the PZT observations by using nearly half common stars in their observations. During the period between 1959 - 1977, results of astronomical observations of both PZTs were obtained [Kakuta *et al.*, 1986] by using the unified catalog of the Washington and Mizusawa PZT stars [Manabe *et al.*, 1984]. The residuals of the ILOM PZT which were derived from the IPMS and the IRIS show good agreements in time and latitude observations [H. Kitago, ILOM, personal communication, 1987]. Here we shall consider local residuals of the astronomical observations to be variations of the vertical at the observation site.

We can show some examples of non-tidal variational relationships of the vertical derived from optical observations as well as those of the geophysical phenomena, such as variations of strain [Kakuta and Sato, 1985], El Niño events [Kakuta, 1986] and secular variations of the *Z* component over Southeast Asia in the 1970's [Mizuno, 1984] and relative variations of time observation residuals between Irkutsk and Japan (Mizusawa and Tokyo) [Kakuta, 1983, 1985].

We shall directly compare couplings between the vertical variations and those of the geomagnetic field. Geomagnetic data is obtained at stations in the west coast of the Pacific, Memanbetsu (43°55'*N*, 144°12'*E*), Kakioka (36°14'*N*, 140°11'*E*), Kanoya (31°25'*N*, 130°53'*E*) and Amberly (43°09'*S*, 172°43'*E*) for *X* (north) and *Y* (east) components.

Variations of the geomagnetic field may be supported by motions in an unstable zone near the inner core boundary (ICB). A gravitationally powered dynamo, which was proposed by Braginsky in 1963, may be a hopeful mechanism for the dynamo theory [Gubbins, 1983; Fearn and Loper, 1985]. There might be a local reversible equilibrium state among the conditions of adiabatic heating and the phase transition due to an increase in pressure for variations over a period of ten years, such as those of the local residuals of astronomical observations.

Lateral heterogeneity in the fluid core is suggested from various points of view. Dziewonsky [1984] showed that a large anomaly of *P* velocity is found in the west coast of the Pacific at the core-mantle boundary (CMB). Bloxham and Gubbins [1987] suggest a predominantly thermal origin for the large variations in seismic velocity observed in the lower mantle. For decade variations heat conduction may be very small and an initial temperature gradient in the horizonal direction may be kept during variations of the local vertical at the station.

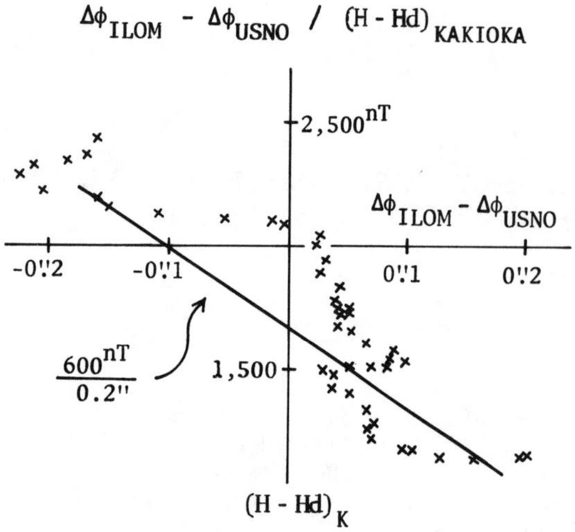

Fig. 1. Correlation between $\Delta\varphi_{ILOM}$-$\Delta\varphi_{USNO}$ and H-Hd at Kakioka.

The purpose of this paper is to study the relationship between local variations of the vertical at the astronomical observation site and those of the geomagnetic field and to discuss motions in the fluid core near the CMB related to variations of the vertical.

Astronomical and Geomagnetic Observations

Differences in the mean latitudes between the ILOM and the USNO are derived from the mean latitude of the ILS VZT at the ILOM [Yumi and Wako, 1966] and the PZT at the USNO [McCarthy, 1972] for the period from 1918 to 1965. By removing the secular pole variation (SV) derived from Vondrák [1985], variations of $\Delta\varphi_{ILOM}$ - $\Delta\varphi_{USNO}$ -(SV) are compared with variations of the horizontal component of the non-dipole field horizontal component *H-Hd*, observed at Kakioka geomagnetic observatory, where *H* stands for the observed value at Kakioka [Kakioka Geomagnetic Observatory, 1983] and *Hd* for the horizontal component derived from the dipole field [Jin and Thomas, 1977]. Figure 1 shows the correlation between $\Delta\varphi_{ILOM}$ - $\Delta\varphi_{USNO}$ and *H-Hd* at the Kakioka geomagnetic observatory.

The astronomical observation results of both PZTs during the period from 1959-1977 [Kakuta *et al.*, 1986] are compared with the annual mean of geomagnetic observations in Japan [Kakioka Magnetic Observatory, 1952-1985, 1983] and the annual mean of $\dot{Y}$ at Amberly, [Gubbins and Tomlinson, 1986]. The local residuals between the ILOM and the USNO and variations of the geomagnetic field are compared by taking a 3 year running mean. Figures 2 and 3 show the differences of $\Delta\varphi$ between the ILOM and the USNO, $\Delta\varphi_{ILOM}$-$\Delta\varphi_{USNO}$ and variations of the *X* component of the geomagnetic field at Kanoya and Memanbetsu. Figures 4 and 5 show $(UT1\text{-}TAI)_{USNO}$ - $(UT1\text{-}TAI)_{ILOM}$

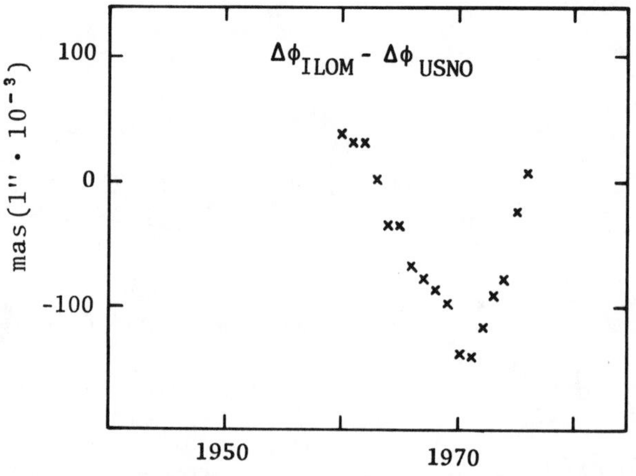

Fig. 2. Differences of $\Delta\varphi$ between ILOM and USNO, $\Delta\varphi_{ILOM}$-$\Delta\varphi_{USNO}$.

and the first time derivative of the geomagnetic field *Y* component between Kakioka and Amberly. Correlations are summarized in Table 1.

Fluid Core Model

In derivations of equations of motion in the fluid core we shall take account of 1)a short time scale variation, 3 ~ 60 years, 2)compressible and non-viscous fluid, 3)the temperature gradient along the meridian, which indicate the isotherm tilt to the potential surface [Kakuta and Onodera, 1972] and 4)motions near the equator with the β- plane method.

We shall define a local coordinate frame as follows; the origin is chosen to coincide with the earth's center of gravity. The z-axis is parallel to the vertical. The x-axis is parallel to the east and the y-axis is taken to the north in the right hand system. $\boldsymbol{I}_\Omega$ is taken to be the unit vector along the mean axis of the earth's

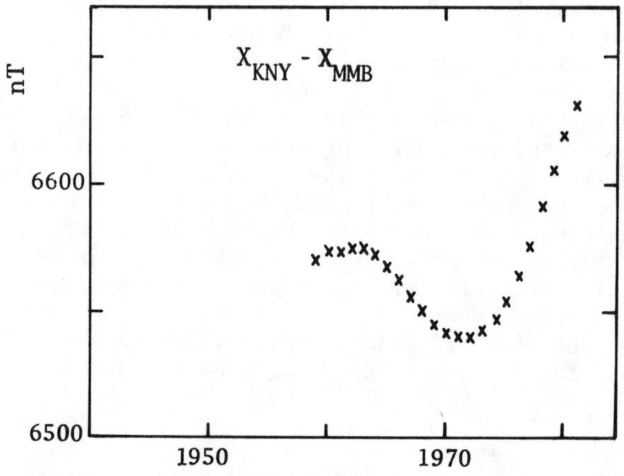

Fig. 3. Variations of X between Kanoya and Memanbetsu, X_{Kanoya}-$X_{Memanbetsu}$.

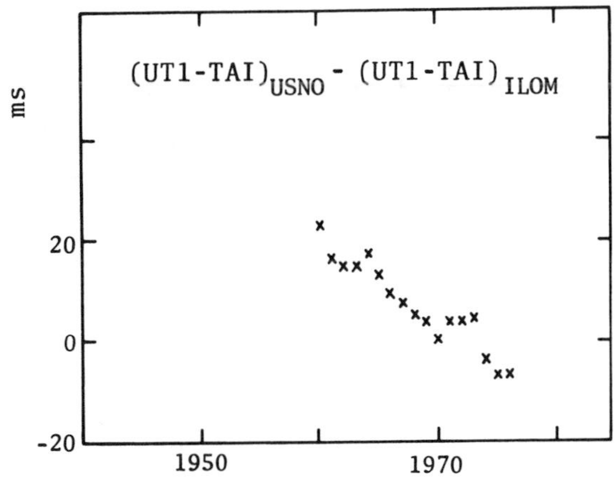

Fig. 4. Differences of UT1 - TAI between USNO and ILOM, $(UT1-TAI)_{USNO}-(UT1-TAI)_{ILOM}$.

TABLE 1. Correlations Between the Astronomical Observations Residuals and Variations in the Geomagnetic Field

Geomagnetic station Component	$\Delta\varphi_{USNO} - \Delta\varphi_{ILOM}$	$(UT1-TAI)_{USNO} - (UT1-TAI)_{ILOM}$
Kakioka H-Hd	$200^{mas}/600nT$	
Memanbetsu-Kanoya X	$78^{mas}/17nT$	
Memanbetsu-Kanoya $\dot{Y}$		$13^{ms}/1.8nT/yr$
Kakioka-Amberley $\dot{Y}$		$10^{ms}/3.5nT/yr$

rotation and close to the unit vector along the y-axis $\boldsymbol{I}_y$.

We can write equations of motion as follows,

$$\rho D^2\boldsymbol{u} + 2\omega\boldsymbol{I}_\Omega \times \boldsymbol{r} = \rho\nabla V + \nabla\tilde{P} \quad (1)$$

$$D = \frac{\partial}{\partial t}$$

and

$$\tilde{p} = -p + \lambda\Delta \quad (2a)$$

$$p = p_0 + \delta p' - \boldsymbol{u}\cdot\nabla p_0 \quad (2b)$$

with

$$\delta\rho = \rho - \rho_0 = -\boldsymbol{u}\cdot\nabla\rho_0 - \rho\Delta \quad (3a)$$

$$\delta T = T - T_0 = -\boldsymbol{u}\cdot\nabla T \quad (3b)$$

$$V = V_0 + K \quad (3c)$$

$$\nabla\rho_0 = \left(\frac{\partial\rho}{\partial p}\right)_T \nabla p_0 + \left(\frac{\partial\rho}{\partial T}\right)_p \nabla T = \frac{\rho}{\lambda}\nabla p_0 - \alpha\rho_0\nabla T \quad (3d)$$

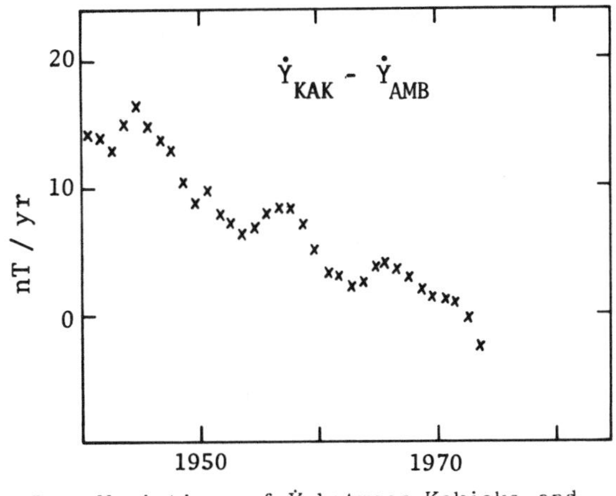

Fig. 5. Variations of $\dot{Y}$ between Kakioka and Amberly, $\dot{Y}_{Kakioka}-\dot{Y}_{Amberly}$.

We shall assume the earth's rotational angular velocity to be expressed by

$$\boldsymbol{\omega} = (0,\ \omega y/r,\ \omega z/r) \quad (4)$$

where $\boldsymbol{u}$ stands for displacement from the hydrostatic equilibrium, ω the magnitude of the Earth's rotational angular velocity, V the geopotential, p the pressure, ρ the density, T the temperature, Δ divergence ($= \nabla\cdot\boldsymbol{u}$), α the coefficient of the thermal expansion and λ the Lamé constant. The notation δ stands for small quantities. The suffix 0 stands for the value at the hydrostatic equilibrium.

The equation of state is derived from Debye's formula [Shimazu, 1966, p. 109]

$$\delta p' = S(\delta\rho T + \rho\delta T) \quad (5a)$$

and

$$S = \frac{3R\gamma}{A} \quad (5b)$$

where R stands for the gas constant, A the mean atomic weight and γ the Grüneisen constant. Here we shall use the parameter indicating the compactness of matter to be the unit as shown in Anderson's [1986] expression (16). The condition of the hydrostatic equilibrium can be expressed by

$$\rho_0\nabla V_0 = \nabla p_0 \quad (6)$$

The Poisson equation is written as,

$$\nabla^2 K = 4\pi f\left\{\left(\frac{\rho}{\lambda}\right)\rho_0\eta - \alpha\rho_0(\boldsymbol{u}\cdot\nabla T) + \rho_0\Delta\right\} \quad (7)$$

and

$$\eta = \boldsymbol{u}\cdot\nabla V_0 \quad (8)$$

By taking the numerical values derived by Stacey [1977], γ=1.419 and A=50 or the fluid core at the CMB, we obtain S=0.708·10^3 m^2 s^{-2} K^{-1}. Variation effects of the pressure due to changes of temperature are taken into consideration and equations of motion (1) are reduced to the following expressions,

$$D^2 \boldsymbol{u} + 2\omega \boldsymbol{I}_\Omega \times D\boldsymbol{u} = -\nabla\Phi + \boldsymbol{u} \times (\nabla V_0 \times \boldsymbol{\beta}) + S_1 \nabla V_0 \Delta \qquad (9)$$

where

$$\Phi = -\eta - \frac{\lambda}{\rho}\Delta - K \qquad (10)$$

$$S_1 = \frac{ST}{\lambda/\rho} - \alpha\frac{\lambda}{\rho}R_T\left(1 - \frac{S}{\alpha\lambda/\rho}\right) \qquad (11a)$$

$$(\boldsymbol{u}\cdot\nabla T) = (\boldsymbol{u}\cdot\nabla V_0)R_T \qquad (11b)$$

$$R_T = \frac{|(\nabla T\cdot\boldsymbol{I}_z)|}{g} \qquad (11c)$$

$$\boldsymbol{\beta} = \alpha\nabla T \qquad (12)$$

and R_T stands for the vertical component ratio of the temperature gradient and $\boldsymbol{\beta}$ the temperature gradient horizontal component. The temperature variation effects on the terms of η are small and neglected in equation (9). The second term, $\boldsymbol{u}\times(\nabla V_0\times\boldsymbol{\beta})$ in the right hand side in equations (9) indicates the isotherm tilt to the potential surface [Kakuta and Onodera, 1972]. We shall define

$$\boldsymbol{\omega}_T = \nabla V_0 \times \boldsymbol{\beta} \qquad (13)$$

and we can rewrite equations (9) as follows,

$$D^2 \boldsymbol{u} + \boldsymbol{F} \times \boldsymbol{u} = -\nabla\Phi + S_1\nabla V_0\Delta \qquad (14a)$$

$$\boldsymbol{F} = 2\omega D\boldsymbol{I}_\Omega + \boldsymbol{\omega}_T \qquad (14b)$$

To solve equation (14a), we shall specify the fluid motion near the equator and

$$\boldsymbol{\beta} = (0,\ \beta_y,\ 0) \qquad (15)$$

$$\nabla V_0 = (0,\ 0,\ -g) \qquad (16)$$

$$\boldsymbol{F} = \lim_{z\to 0}\left(g\beta_y,\ 2\omega Dy/r,\ 2\omega Dz/r\right) \qquad (17)$$

and

$$\nabla\cdot\boldsymbol{F} = 2\omega D/r \qquad (18)$$

The expression (18) will be kept in the following analysis in order to consider the β-plane method.

For a slow variation we can reduce equations (14a) to

$$D^2(\boldsymbol{F}\cdot\boldsymbol{F})\boldsymbol{u} = D^2(\boldsymbol{F}\times\nabla\Phi) - \boldsymbol{F}(\boldsymbol{F}\cdot\nabla)\Phi + D^2[gS_1(\boldsymbol{F}\times\boldsymbol{I}_z)\Delta] \qquad (19)$$

Equations of η, Δ and K are derived from equations (19) and (7) with the aid of (18).

$$\begin{bmatrix} D^2(\boldsymbol{F}\cdot\boldsymbol{F}) & 0 & 0 \\ 0 & D^2[(\boldsymbol{F}\cdot\boldsymbol{F})+gS_1Q_z] & 0 \\ 0 & -D^2l^2S_1 & D^2(\boldsymbol{F}\cdot\boldsymbol{F})\left\{\dfrac{D^2}{V_g^2}+\dfrac{1}{C_v^2}\right\} \end{bmatrix}\begin{bmatrix} \eta \\ \Delta \\ K \end{bmatrix}$$

$$= -\begin{bmatrix} gD^2Q_z \\ (\boldsymbol{F}\cdot\nabla)\left\{(\boldsymbol{F}\cdot\nabla)+\dfrac{2\omega D}{r}\right\} \\ D^2\left\{\dfrac{(\boldsymbol{F}\cdot\boldsymbol{F})}{c_v^2}-\dfrac{l^2}{g}\nabla_z\right\}-2\omega D\beta_y \end{bmatrix}\Phi \qquad (20)$$

where

$$l = g\beta_y \qquad (21a)$$

$$Q_z = F_x\nabla_y - F_y\nabla_x \qquad (21b)$$

$$\nu_g^2 = 4\pi f\rho \qquad (21c)$$

$$c_v^2 = \lambda/\rho \qquad (21d)$$

With the aid of equation (10) we obtain equation of Φ from equations (20) as follows,

$$\left\{\frac{\nabla^2}{\nu_g^2}+\frac{1}{c_v^2}\right\}\left[\left\{(\boldsymbol{F}\cdot\boldsymbol{F})+gS_1Q_z\right\}gD^2Q_z+c_v^2(\boldsymbol{F}\cdot\boldsymbol{F})(\boldsymbol{F}\cdot\nabla)\right.$$

$$\left.\cdot\left\{(\boldsymbol{F}\cdot\nabla)+\frac{2\omega D}{r}\right\}-D^2(\boldsymbol{F}\cdot\boldsymbol{F})\left\{(\boldsymbol{F}\cdot\boldsymbol{F})+gS_1Q_z\right\}\right]\Phi$$

$$+\left[\left\{(\boldsymbol{F}\cdot\boldsymbol{F})+gS_1Q_z\right\}\left[\left\{\frac{(\boldsymbol{F}\cdot\boldsymbol{F})}{c_v^2}-\frac{l^2}{g}\nabla_z\right\}-2\omega D\beta_y(\boldsymbol{F}\cdot\nabla)\right]\right.$$

$$\left. +S_1l^2(\boldsymbol{F}\cdot\nabla)+\frac{2\omega D}{r}\right]\Phi = 0 \qquad (22)$$

Discussion

We shall consider a slow variation in the fluid core. Numerical values for quantities at the CMB included in equation (22) are taken as follows,

$$g=1.068\ 23\cdot10^1\ m\ s^{-2} \qquad \rho=9.903\ 49\cdot10^3\ kg\ m^{-3}$$

[Dziewonski and Anderson, 1981]

$$T=3.157\cdot10^3\ K \qquad \alpha=1.57\cdot10^{-5}\ K^{-1}$$

[Stacey, 1977]

$$\lambda/\rho=K_T/\rho \qquad K_T=6.522\cdot10^{11}\ Pa$$

[Stacey, 1969, p. 282]

and

$$S_1 = 3.4\cdot10^{-2}$$

$$\nu_g^2 = 4\pi f\rho = 1.5\cdot10^{-5}\ s^{-2}$$

To solve (22) we assume the form of Φ to be as

$$\Phi = \Phi_0\{i(k_i\chi_i + \sigma t)\} \qquad i = x, y, z \tag{23}$$

We assume the magnitude of the wave number k_i to be $|k_x|<|k_y|<|k_z|$, and a typical value for β_y as $|\beta_y|=1\cdot 10^{-13}$.

Approximate expression of equation (22) for a slow variation is written as

$$\left(\frac{\partial^2}{\partial x^2} + \frac{2\omega}{rl}D\frac{1}{\Gamma_K}\frac{\partial}{\partial x}\right)\Phi = 0 \tag{24}$$

The characteristic frequency, which is defined in expression (23), is obtained from equation (24) as,

$$\sigma = -\frac{rg\beta_y k_x}{2\omega}\Gamma_K$$

and

$$\Gamma_K = \left(1+S_1-\frac{k_z^2}{\nu_g^2}c_v^2\right)/\left(1+S_1-r\beta_y-\frac{k_z^2}{\nu_g^2}c_v^2\right) \tag{25}$$

where r stands for the radius of the fluid core at the CMB.

By using expression (23) with characteristic frequency σ in expression (24), we obtain η, Δ and K from expressions (20) and (10)

$$\eta = -i\frac{k_y}{\beta_y}\Phi \tag{26a}$$

$$\Delta = \left(\frac{2\omega}{g}\right)^2\frac{1}{r\beta_y\Gamma_K^2}\Phi \tag{26b}$$

$$K = \left\{1-\left(\frac{2\omega}{g}\right)^2\frac{c_v^2}{r\beta_y\Gamma_K^2} + i\frac{k_y}{\beta_y}\right\}\Phi \tag{26c}$$

If we take $|k_x|=|k_y|=|\beta_y|$, we obtain $|\sigma|\sim 10^{-10}\mathrm{sec}^{-1}$, $|\eta| = |\Phi|$, and $|K|\sim|(\lambda/\rho)\Delta|\sim 10^2|\Phi|$.

Compressibility effects on a wave motion may be important. Compressible fluid motion might relate to the gravitationally powered dynamo in a local equilibrium state. The wave motion is derived from an approximation of the β- plane treatment and a Rossby type wave and propagates in an east-west direction. The direction of propagation depends on the sign of β_y.

The deflections of the vertical (VD) are expressed as follows

$$VD = \nabla K/g$$

The relation between the geopotential and the deflections of the vertical is shown as

$$|VD| = (|rk_x|/2)\cdot|K/V_0|$$

Suppose $|VD|=1$ $mas=0.5\cdot 10^{-8}$, the magnitude of the disturbing potential is

$$|K| = 10^{-8}\cdot|V_0/(rk_x)|$$

Variations of the disturbing potential may contribute to variations in the geomagnetic field and the geopotential in the global scale. Abrupt contour changes of $\dot{Z}=0$ in the 1970's over Southeast Asia [Mizuno, 1984] and depressions of sea level at Truk Island in 1972 [Kakuta, 1986] occurred near the Western Pacific in the early period of the 1970's.

Conclusion

Relationships between the optical observations of time and latitude residuals, which are derived from the ILOM and the USNO, and local variations of the geomagnetic field are studied.

The latitude observation residuals show a correlation with local horizontal component residuals and the X component of the geomagnetic field. The time observation residuals, however, show a correlation with local variations of the first time derivative in the Y component.

A possible explanation of the correlation between variations of the vertical and those of the geomagnetic field near the CMB is attempted by taking into account the meridional component of the temperature gradient in a non-viscous, compressible and rotating fluid near the CMB. The possibility of a Rossby type wave to induce variations of the disturbing potential as well as those of the vertical near the CMB, is found. The results show that the disturbing potential should be large enough to explain variations of the vertical.

Acknowledgments. The author gratefully acknowledges the aid of Yuko Oouchi in the preparation of this manuscript.

References

Anderson, O. L., Properties of iron at the earth's core conditions, Geophys. J. R. Astron. Soc., 84, 561-579, 1986.

Bloxham, J., and D. Gubbins, Thermal core-mantle interactions, Nature, 325, 511-513, 1987.

Dziewonski, A. M., Mapping the lower mantle: determination of lateral heterogeneity in P velocity up to degree and order 6, J. Geophys. Res., 89, 5929-5952, 1984.

Dziewonski, A. M., and D. L. Anderson, Preliminary reference earth model, Phys. Earth Planet. Inter., 25, 297-356, 1981.

Fearn, D. R., and D. E. Loper, Pressure freezing of

the earth's inner core, Phys. Earth Planet. Inter., 39, 5-13, 1985.

Gubbins, D., The influence of extrinsic pressure changes on the earth's dynamo, Phys. Earth Planet. Inter., 33, 255-259, 1983.

Gubbins, D., and L. Tomlinson, Secular variation from monthly mean from Apia and Amberley magnetic observatories, Geophys. J. R. Astron. Soc., 86, 603-616, 1986.

Jin, R. S., and D. M. Thomas, Spectral line similarity in the geomagnetic dipole field variations and length of day fluctuation, J. Geophys. Res., 82, 828-834, 1977.

Kakioka Magnetic Observatory, Report of the Kakioka Magnetic Observatory, Geomagnetism, Kakioka, Memambetsu and Kanoya, 1952-1985.

Kakioka Geomagnetic Observatory, Centenary History of Geomagnetic Observation, (in Japanese), 167, 1983.

Kakuta, C., Variations of longitude and latitude in the Southeast Asian plate, J. Geod. Soc. Japan, 29, 223-235, 1983.

Kakuta, C., Local residuals of astronomical observations of longitude and latitude and motions near the core-mantle boundary, J. Geod. Soc. Japan, 31, 261-272, 1985.

Kakuta, C., Variations of longitude differences and long term variations of the ocean, Publ. Int. Latit. Obs. Mizusawa, 19, 15-28, 1986.

Kakuta, C., and E. Onodera, Global oscillation of the ocean supported by geothermal energy flow, Publ. Int. Latit. Obs. Mizusawa, 8, 119-126, 1972.

Kakuta, C., and K. Sato, Variations of strain and changes of the earth's rotation, J. Geod. Soc. Japan, 31, 163-176, 1985.

Kakuta, C., D. D. McCarthy, T. Hara, K. Sato, K. Yokoyama, S. Manabe, S. Sakai, H. Kitago, K. Iwadate, A. K. Babcock, I. W. Lindenblad, and L. Hinnov, Non-tidal motions between Mizusawa and Washington, D. C. with the aid of the PZT observations, Publ. Int. Latit. Obs. Mizusawa, 19, 1-13, 1986.

Manabe, S., S. Sakai, T. Hara, H. Kitago, and K. Iwadate, Unified Catalog of Washington and Mizusawa PZT stars, Publ. Int. Latit. Obs. Mizusawa, 63-92, 1984.

McCarthy, D. D., Secular and non polar radiation of Washington latitude, in Rotation of the Earth, edited by P. Melchior and S. Yumi, pp. 86-96, Reidel, Dordrecht, 1972.

Mizuno, H., Rapid and episodic variation of the geomagnetic secular variation field, Bull. Geographical Survey Institute, 29, 1-102, 1984.

Shimazu, Y., Chikyu Naibu Butsuri Gaku, (Physics of the Earth's Interior)(in Japanese), 394 pp., Shokabo, Tokyo, 1966.

Stacey, F. D., Physics of the Earth, 324 pp., John Wiley, New York, 1969.

Stacey, F. D., A thermal model of the earth, Phys. Earth Planet. Inter., 15, 341-348, 1977.

Vondrák, J., Long-period behaviour of polar motion between 1900.0 and 1984.0, Annales Geophysicae, 3, 351-356, 1985.

Yumi, S., and Y. Wako, On the secular motion of the mean pole, Publ. Int. Latit. Obs. Mizusawa, 5, 31-86, 1966.

DYNAMICS OF THE EARTH'S CORE AND THE GEODYNAMO

F. H. Busse and K.-K. Zhang

Institute of Physics, University of Bayreuth, 8580 Bayreuth, West-Germany,
and Institute of Geophysics and Planetary Physics,
University of California at Los Angeles

Abstract. Current ideas related to the problem of the generation of the geomagnetic field by motions in the liquid iron core of the Earth are reviewed and the difficulties of theoretical approaches are outlined. The importance of theoretical models starting from basic principles is emphasized and recent results from computations by Zhang and Busse [1988] of the magnetohydrodynamic problem of magnetic field generation by buoyancy driven convection in rotating spherical shells are described. This particular approach follows the sequence of bifurcating solutions of the full set of nonlinear equations. Future extensions and possible limitations of this numerical approach are discussed in a concluding section.

Introduction

The liquid core of the Earth is not only one of the most inaccessible places of the universe for scientific inquiry, but the very data that could provide information about its dynamics are also the ones for which the interpretation is most difficult. Thus after centuries of data collection on the variations in time of the geomagnetic field we are just at the beginning of an understanding of these data on the basis of reasonably reliable theoretical models. However, the progress in the general theory of the dynamo process and the increasing speed and capacity of computers give rise to optimism. Before discussing recent progress and future directions of the field, it seems appropriate to describe the major obstacles and open questions in understanding of the geodynamo. The following list is far from complete, but may be helpful in stimulating new thoughts about the problem.

(i) Knowledge about the material properties of the core is sparse. Important properties such as the kinematic viscosity are not known even within an order of magnitude. The question whether parts of the core have stably stratified density distributions has not yet been settled. Inversions of improved seismic data could help to answer this question. The condition that unstably stratified regions are physically unrealistic should be used as a side constraint in those inversions.

(ii) Compositional buoyancy owing to light elements left in solution by the solidification of iron at the solid inner core is considered the most likely source of mechanical energy. But the way in which this source of energy becomes available to drive large scale motions in the outer core is not well understood. The influence of postulated "mushy zones" [Loper and Roberts, 1982; Roberts and Loper, 1982] on the dynamics of the core may be significant.

(iii) The possible use of geomagnetic secular variation data for the determination of the velocity fields at the surface of the core has been a subject of intense research in recent years. But this research offers little information on those components of the magnetic field and the velocity field which are screened by the finite conductivity of the lower mantle because they are of too small scale either in time or in space.

(iv) Any magnetic field in the core can be separated into a poloidal and a toroidal part of which only the former emerges from the conducting core in the form of the potential field which can be measured at the Earth's surface. Although it is generally believed that the toroidal field is at least of the same order of magnitude as the poloidal field in the core, opinions about its actual strength vary widely.

(v) The apparent azimuthal inhomogeneity of the dynamo process (low secular variation in the Pacific hemisphere, high secular variation in Atlantik hemisphere as recently emphasized by Bloxham and Gubbins [1985, 1987], and its possible connection with processes in the Earth's mantle complicate the theoretical understanding of core dynamics because of the additional parameters that are introduced. On the other hand, correlation with other geophysical observables may eventually help to discriminate between various models of the geodynamo.

(vi) The aperiodically occuring reversals certainly represent the most fascinating property of the geodynamo. Despite numerous speculations very little is known about the origin of reversals. Because of the small ratio between the time span of the reversal process and the mean recurrence time of reversals, their simulation in numerical models poses special challenges. Ultimately the increasing amount of paleomagnetic data on reversals will provide the most crucial tests for theories of the geodynamo.

This list of difficulties and questions could easily be continued, especially since the discovery of the magnetic fields of several planets has opened up new possible applications for theories of the dynamics in rotating spherical shells. But instead of continuing this list we shall turn to the formulation of mathematical approaches.

Mathematical Foundations of the Problem

Because of the many uncertain external parameters for the problem of the geodynamo it is prudent to proceed with a model featuring the smallest number of parameters while including all physical effects that may have a significant influence on the operation of the geodynamo. Obviously judgments of this matter are subjective and different choices will be made for different goals of the subsequent analysis. Here we like to start with the following equations for the velocity vector $\underset{\sim}{u}$, the flux density $\underset{\sim}{B}$, and the deviations Θ and Σ of the temperature and of the concentration of light elements from the spherically symmetric static solution of the problem,

$$\left(\frac{\partial}{\partial t}+\underset{\sim}{u}\cdot\nabla\right)\underset{\sim}{u}+2\underset{\sim}{\Omega}\times\underset{\sim}{u} = -\nabla\pi-\underset{\sim}{g}(\alpha\Theta+\delta\Sigma)+\nu\nabla^2\underset{\sim}{u}+\frac{1}{\rho_0\mu}(\nabla\times\underset{\sim}{B})\times\underset{\sim}{B} \quad (1)$$

$$\nabla\cdot\underset{\sim}{u} = 0 \quad (2)$$

$$\left(\frac{\partial}{\partial t}+\underset{\sim}{u}\cdot\nabla\right)\Theta+\underset{\sim}{u}\cdot\nabla T_s = \kappa\nabla^2\Theta \quad (3)$$

$$\left(\frac{\partial}{\partial t}+\underset{\sim}{u}\cdot\nabla\right)\Sigma+\underset{\sim}{u}\cdot\nabla S = D\nabla^2\Sigma \quad (4)$$

$$\left(\frac{\partial}{\partial t}+\underset{\sim}{u}\cdot\nabla\right)\underset{\sim}{B} = \underset{\sim}{B}\cdot\nabla\underset{\sim}{u}+\lambda\nabla^2\underset{\sim}{B} \quad (5)$$

where ν,κ,λ,D denote the kinematic viscosity, the thermal diffusivity, the magnetic diffusivity, and the diffusion constant for the light elements. We have assumed the Boussinesq approximation, in which the variability of the density

$$\rho = \rho_0(1-\alpha(T-T_0+\Theta)-\delta(S-S_0+\Sigma)) \quad (6)$$

is taken into account only in connection with the gravity term. All terms that can be written in the form of a gradient in equation (1) have been combined in $\nabla\pi$. In writing the heat equation (2) we have allowed, however, for effects of compressibility in that the property has been taken into account that only the difference T_s between the actual spherically symmetric temperature distribution $T(r)$ and the adiabatic temperature distribution $T_{ad}(r)$ can give rise to deviations Θ of the temperature owing to advection. The effects of Ohmic heating are neglected in accordance with the Boussinesq approximation [see, for example, Gray and Giorgini, 1976]. Because the characteristic scale height of the core fluid is much larger than the core radius, equations of the form (1-5) are well suited for an approximate description of the dynamics of the core. For similar reasons all material properties are assumed to be constant.

The similarity of the appearance of the variables Θ and Σ in equations (1-5) suggests that one of the two variables can be neglected in the theory in first approximation. This idea depends, of course, on the signs and on the relative magnitudes of the radial gradients of the function $S(r)$ and $T_s(r)$, since problems such as thermohaline convection in the oceans provide examples where Θ and Σ play very different roles. In the present context we are interested, however, only in large scale motions and it seems reasonable to drop Σ from the problem and interprete instead Θ either as temperature deviation or as the concentration deviation or as a combination of both. These interpretations may require, of course, an appropriate adjustment of the boundary conditions and of the definition of κ.

The geometrical configuration of the problem is shown in figure 1. The ellipticities of the core-mantle boundary of the inner core are neglected and no topographic effects on those boundaries are considered. A spherical system of coordinates is typically used in the analysis of the problem. The force of gravity can be assumed in the form $\underset{\sim}{g} = -\gamma\underset{\sim}{r}$ where $\underset{\sim}{r}$ is the dimensionless position vector based on the length scale r_0-r_i; r_0 and r_i denote the radii of the inner and outer core, respectively, and $\eta = r_i/r_0$ is the radius ratio. There exist few arguments which could serve as guides in selecting geophysically realistic functions $T_s(r)$ or $S(r)$. If, for instance, in an adiabatically stratified core the first function vanishes, the process of liquid enriched in the light elements rising through the outer core can be compared with convection in a fluid layer thermally insulated from above and

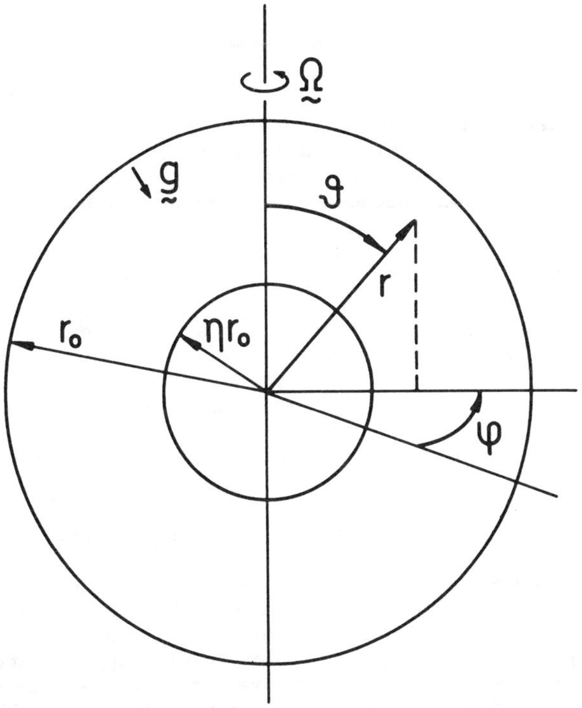

Fig. 1. Geometrical configuration of the rotating spherical fluid shell

heated from below. In addition, a homogeneous distribution of cooling sources in the fluid could be assumed in order to obtain a steady state process. The corresponding function S(r) would vary strongly near the inner-outer core boundary and attain a vanishing slope at the core-mantle boundary. Traditionally, however, the problem of thermal convection in a rotating spherical shell has been studied for a homogeneously heated sphere corresponding to $T_s(r) = T_0 - \beta r^2$. Since the qualitative features of convection do not seem to be affected much by the particular choice for $T_s(r)$ and since the equations assume a particularly simple form for the traditional choice, this choice has been retained for the computations to be reported in the following.

The mathematical problem can be simplifid by the introduction of the general representations

$$\underset{\sim}{u} = \frac{\kappa}{r_0 - r_i}\,[\nabla\times(\nabla\times\underset{\sim}{r}\Phi)+\nabla\times\underset{\sim}{r}\Psi] \qquad (7)$$

$$\underset{\sim}{B} = [\rho_0\mu\kappa^2/(r_0-r_i)^2]^{1/2}[\nabla\times(\nabla\times\underset{\sim}{r}h)+\nabla\times\underset{\sim}{r}g)] \qquad (8)$$

for the solenoidal vector fields $\underset{\sim}{u}$ and $\underset{\sim}{B}$. The factors in front of the brackets containing the poloidal and toroidal components of the vector fields have been chosen in such a way that the scalar functions Φ,Ψ,h, and g are dimensionless. By taking the r-components of the curl and the $(\text{curl})^2$ of equation (1) and by taking the r-components of equation (1) and of its curl, five equations for the five unknowns Φ,Ψ,h,g, and Θ can be obtained. The number of dependent variables has thus been decreased considerably albeit at the expense of an increase in the order of the differential operators in the equations.

Since in the limit of high rotation rates the solutions for stress-free and no-slip boundary conditions for the velocity field tend to approach each other, the simpler stress-free conditions are usually assumed in numerical work. To avoid separate computations for the magnetic field outside the fluid outer core either a vanishing or an infinite electrical conductivity must be assumed in those regions. The boundary conditions given below apply in the former case. More realistic boundary conditions are desirable, especially if effects such as electromagnetic core-mantle coupling are studied. Such extensious of the present work will have high priority in the future. Finally the boundary condition for Θ must be specified. The vanishing of Θ at the inner boundary and of its radial derivative at the outer boundary would be appropriate choices. Again for reasons of tradition we have adopted instead in the present work a vanishing Θ at the outer boundary

$$\Phi = \frac{\partial^2}{\partial r^2}\Phi = \frac{\partial}{\partial r}\frac{\Psi}{r} = \Theta = g = 0 \text{ at } r = \eta(1-\eta)^{-1},\ (1-\eta)^{-1}\ . \qquad (9)$$

The boundary condition for h results from the conditions for the matching with the potential field outside the fluid shell.

There is no need at this point to write down more equations. For later use we just mention the four dimensionless parameters which incorporate all material properties and the external parameters of the system,

Rayleigh number	$R = \alpha\gamma\beta(r_0-r_i)^3/\nu\kappa$
Taylor number	$T = 4\Omega^2(r_0-r_i)^4/\nu^2$
Prandtl number	$P = \nu/\kappa$
magnetic Prandtl number	$P_m = \nu/\lambda$

Symmetry Considerations and Bifurcation Sequences

Details on the numerical methods for the solution of the equations for Φ,Ψ,Θ,h,g are given in a sequence of papers [Zhang and Busse, 1987, 1988, work to be submitted to publication 1988]

which are either published or to be published in the near future. Here we like to focus the attention on some general aspects of the manifold of solutions.

While the trivial static solution of the problem,

$$\Phi \equiv \Psi \equiv \Theta \equiv h \equiv g \equiv 0 \tag{10}$$

exists for all parameter values, non-trivial solutions can exist only for values of R exceeding a value R_s which depends on T and P. In some region of the parameter space the value R_s is identical with critical value R_c determined from the linearized equations for Φ, Ψ, and Θ. Among the manifold of solutions that have been investigated, solutions which are periodic in the azimuthal direction and feature functions Φ, Θ that are symmetric with respect to the equatorial plane are physically preferred since they correspond to the critical value of the Rayleigh number. Although the function Ψ has an symmetry opposite to that of Φ, we shall call those solutions symmetric. There exist another set of solutions with opposite symmetry for the components whose periodicity is given by the basic azimuthal wave number m_0, but these solutions correspond to much higher Rayleigh numbers except in special cases [see, for example, Geiger and Busse, 1981]. All solutions that are not axissymmetric have the property that they are time dependent, in general, for finite values of T. For the symmetric solutions bifurcating from the static state (10) at $R = R_c$ this time dependence assumes the form of a drift of the convection pattern in the azimuthal direction. It thus becomes possible to describe these solutions as stationary solutions with respect to a drifting frame of reference.

The symmetric solution starting at $R = R_c$ is characterized by a vanishing magnetic field, $h \equiv g \equiv 0$, since equation (5) requires a finite magnitude of the velocity field for nondecaying solutions. As the symmetric solution grows in amplitude with increasing Rayleigh number, it usually becomes unstable and a secondary solution branches off. The growing disturbances can be either of the hydrodynamic type with finite perturbations $\tilde{\Phi}, \tilde{\Psi}, \tilde{\Theta}$ and vanishing h,g or of the magnetic type with finite h,g, in which case we speak of a dynamo. As the magnetic instability grows it also generates perturbations of the velocity field through the action of the Lorentz force. But those perturbations grow only proportionally to the square of the amplitude of the magnetic field.

Because of the symmetry of the primary solution the secondary dynamo solutions can be separated into two classes

dipolar class:	h is antisymmetric,	
	g is symmetric	(11)
quadrupolar class:	h is symmetric,	
	g is antisymmetric	(12)

where the symmetry refers to the equatorial plane. Once the secondary solution grows the symmetry can be changed only by further bifurcations because the symmetry of the Lorentz force is the same for both classes (11-12). The fact that bifurcations corresponding to the two classes (11-12) of the magnetic field are not clearly separated in the parameter space indicates the possibility of a close competition between two different dynamos and may provide the key for the understanding of excursions and reversals of the geomagnetic field. There are, however, some differences in the nonlinear properties of the two classes as we shall discuss below.

Some Results of Recent Computations

Numerical computations based on equations (1-5) have been carried out for a number of years. In the earlier work of Cuong and Busse [1981] the axisymmetric interaction approximation had been introduced to overcome the restriction of limited available computer capacity. This approximation allowed the neglection of higher harmonics in the azimuthal direction. As nonlinear effects become important, however, the assumptions on which the approximation is based turn out to be less well satisfied than had been expected and significant differences are seen in the comparison with the full solution as shown by Bolton [1985] in the case without magnetic field. Here we like to mention some results of recent computations [Zhang, 1987; Zhang and Busse, 1987, 1988] in which the number of coefficients in the Galerkin representations of the five fields Φ, Ψ, Θ,h,g has been increased sufficiently, that a reliable approximation for the exact solution can be obtained. Some typical results are shown in figure 2. It is worth noting that numerical computations of spherical dynamos based on equations similar to equation (1-5) have been performed by Gilman [1983] and Glatzmaier [1984] in order to model the solar magnetic cycle. Since an oscillatory dynamo was investigated and because different boundary conditions and different numerical methods were employed, a direct comparison with the results is not possible.

The following list of some results of general nature is based on computations in a limited parameter space. For details we refer to our above mentioned papers.

(i) The dynamo process occurs primarily outside the cylindrical surface touching the inner core at its equator. This property reflects the fact that convection outside this surface is possible at much lower Rayleigh numbers than inside the surface [Busse and Cuong, 1979].

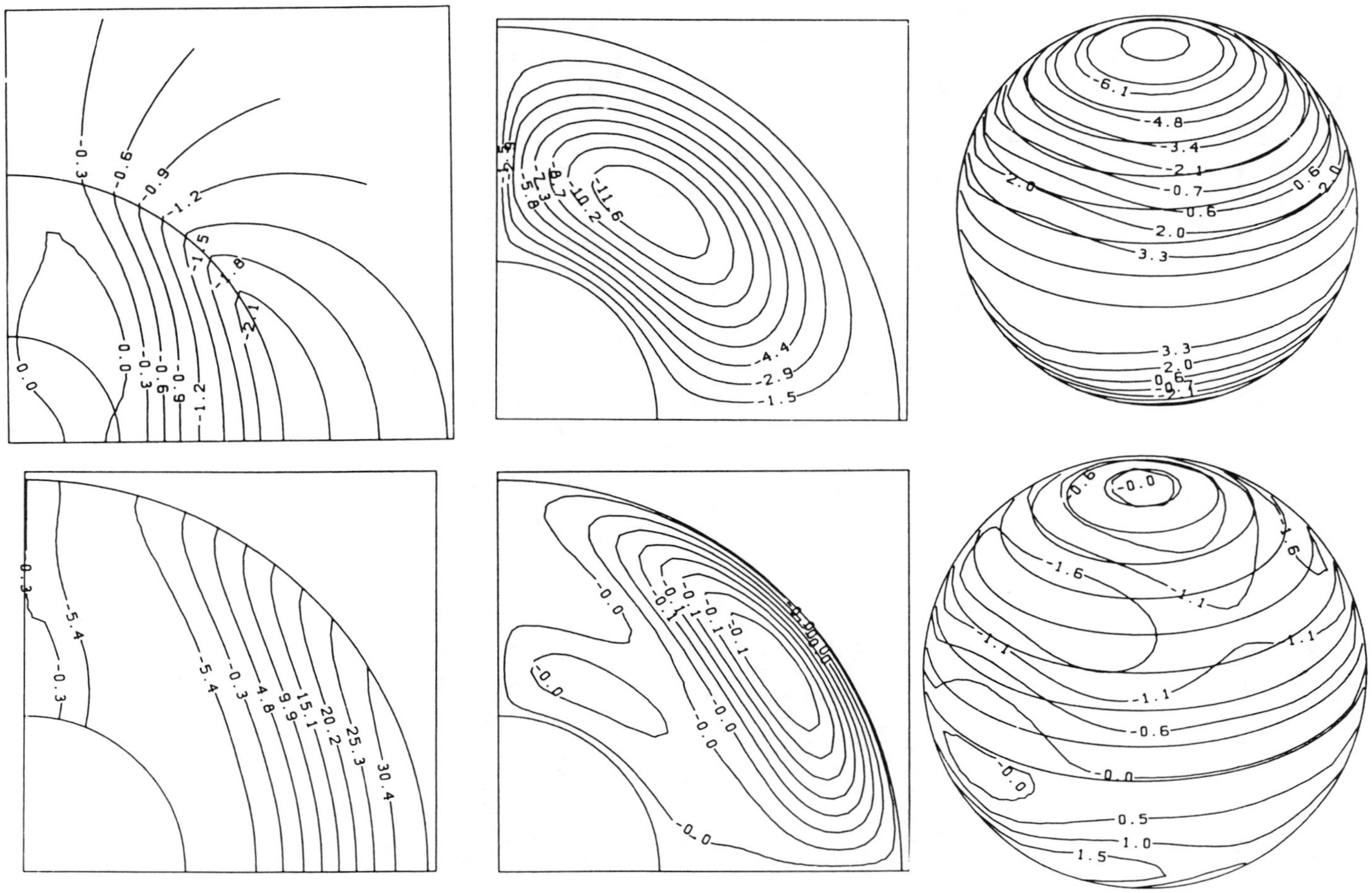

Fig. 2. Clockwise from upper left, field lines of mean poloidal magnetic field, lines of constant mean toroidal field strength, toroidal magnetic field lines at mid-radius, lines of constant radial component B_r at the outer radius, stream lines of meridional circulation, lines of constant angular velocity, all at R = 4000, T = 58700, P = 0.1, P_m = 14.7, η = 0.4 for an m = 2-periodic solution of the dipolar class.

(ii) Typically the onset of dynamo action occurs in the form of a subcritical bifurcation and the amplitude of convection in the presence of the magnetic field is significantly larger than in its absence. This feature has been expected since the basic physical reason a for planetary magnetic field is believed to be the release of the constraint of rotation by the action of the Lorentz force. General arguments [Chandrasekhar, 1961; Eltayeb and Roberts, 1970] suggest that an Elsasser number of the order unity is required for the optimal balance between Coriolis and Lorentz force. The dynamo computations do indeed indicate that the magnetic energy saturates when the Elsasser number

$$\Lambda = \frac{B_0^2}{2\Omega\rho\mu\lambda} \quad (13)$$

reaches a value of the order unity.

(iii) Using a typical value of $\sigma = 6\cdot10^5 \text{mhom}^{-1}$ for the electrical conductivity of the liquid outer core, Λ attains a value of the order unity for a toroidal field strength exceeding that of the poloidal field by a factor of about 3. Similar ratios between amplitudes of the toroidal and poloidal axisymmetric components of the magnetic field have been found in the computational solutions. But the results vary widely in dependence on the parameters of the problem and in some cases it has been found that the poloidal field is actually stronger than the toroidal field.

(iv) The amplification of the axisymmetric toroidal component of the magnetic field arises both through the fluctuating components of the velocity field and through the differential rotation component with the proportion depending on the parameters of the problem. In other words, the dynamo is both an α^2-dynamo and $\alpha\omega$-dynamo with latter dominating in the case of higher Taylor numbers. In accordance with its amplying influence, the differential rotation tends to be opposed by the Lorentz force. Even though the amplitude of convection is always increased in the presence of the magnetic field, the mean zonal flow component is sometimes diminished.

(v) Critical magnetic Reynolds number for dynamos of both classes (11-12) are of the same order of magnitude and there is no clear preference for the dipolar class from the kinematic computations which give the bifurcation point of the dynamo solutions. There appears to be a significant difference, however, in the interaction of the magnetic field with the mean zonal flow. Since the differential rotation tends to be nearly constant on cylinders coaxial with the axis of rotation, the meridional field lines of the dipolar field can align themselves with the surfaces of constant angular velocity of the zonal flow, while the corresponding fields of the quadrupolar field do not have this possibility. Accordingly the inhibiting influence of the differential rotation is much stronger in the quadrupolar case, and quadrupolar dynamos exhibit smaller mean zonal flows than dipolar dynamos.

Concluding Remarks and Outlook

The computations of which some results have been summarized here are continuing at the present time and the dependence of the solutions on the parameters of the problem will become clearer after more computational results have been analyzed. The stability analysis of the steadily drifting solutions with and without magnetic field has not been mentioned in this paper although it represents an important aspect of the project. The subsequent bifurcation of solutions with two characteristic frequencies will require integration in time of the basic equations. Computations of this kind are presently in progress. The stability analysis also indicates at which points bifurcations of solutions with different symmetries, especially in the case of the magnetic field, may occur. The computations of solutions with lesser symmetry increases the demands on computer capacity; but those computations can easily be accomodated on todays supercomputer.

As we have indicated in the introduction, the mayor difficulty in understanding the geodynamo does not arise from insufficient computer power. The numerous barely recognized parameters and interactions that contribute to the complexity of the dynamics of the Earth's core can be explored in a step by step fashion in conjunction with research in seismology, high pressure physics and other fields. The theoretical approach outlined in this paper will be useful in identifying the most sensitive parameters and by representing the variety of possible dynamical states of the core. Claims for realistic simulations of the core dynamics would be misleading and are certainly not intended at the present state of the research.

Acknowledgment. The research by the authors has been supported by the Deutsche Forschungsgemeinschaft and by the Geophysics Section of the U.S. National Foundation.

References

Bloxham, J. and Gubbins, D., Nature 317, 777-781, 1985.

Bloxham, J. and Gubbins, D., Nature 325, 511-513, 1987.

Bolton, E.W., PhD. thesis, University of California at Los Angeles, 1985.

Chandrasekhar, S., Hydrodynamic and Hydromagnetic Stability, Clarendon Press, Oxford, 1961.

Cuong, P.G. and Busse, F.H., Phys. Earth Planet. Inter. 24, 272-283, 1981.

Eltayeb, I.A. and Roberts, P.H., Astrophys. J. 162, 699-701, 1970.

Geiger, G. and Busse, F.H., Geophys. Astrophys. Fluid Dyn. 18, 147-156, 1981.

Gilman, P.A., Astrophys. J. Suppl. 53, 243-268, 1983.

Glatzmaier, G.A., J. Comp. Phys. 55, 461-484, 1984.

Gray, D.D. and Giorgini, A., Int. J. Heat Mass Transfer 19, 545-551, 1976.

Loper, D.E. and Roberts, P.H., pp. 297-327 in Stellar and Planetary Magnetism (ed. A.M. Soward) Gordon and Breach, 1982.

Roberts, P.H. and Loper, D.E., pp. 329-349, in Stellar and Planetary Magnetism, (ed. A.M. Soward) Gordon & Breach, New York, London, Paris, 1982.

Zhang, K.-K., Ph.D. thesis, University of California at Los Angeles, 1987.

Zhang, K.-K. and Busse, F.H., Geophys. Astrophys. Fluid Dyn. 39, 119-147, 1987.

Zhang, K.-K. and Busse, F.H., Geophys. Astrophys. Fluid Dyn., in press 1988.

AN INCLINED $\alpha^2\omega$ DYNAMO

Anthony M. K. Szeto

Department of Earth and Atmospheric Science
York University, North York, Canada M3J 1P3

Abstract. A model of the Earth's deep interior has previously been proposed (Szeto & Smylie 1984a,b) where the inner core is inclined to the rest of the Earth. That model is a mechanical one involving no MHD dynamo action. For it to be valid the fluid core must be capable of producing, or maintaining against ohmic decay, a magnetic field which is inclined to the rotation axis of the bulk of the Earth. In this paper we develop a numerical scheme for a kinematic dynamo incorporating two boundaries around the fluid core (namely the core–mantle and inner core–cuter core interfaces), differential rotation at both boundaries, and the α effect. Assumption of a spatially constant α allows a straight forward analysis in terms of spherical Bessel functions, resulting in an eigenvalue problem for the critical magnetic Reynolds number. This work represents the first effort in constructing an $\alpha^2\omega$ dynamo in a "realistic" geometry, which is an extension of a recent calculation by St Pierre (1987).

Introduction

Several years ago Szeto and Smylie (1984b) began to investigate the feasibility of an inclined inner core, a situation where the symmetry and rotation axes of the inner core are assumed to deviate significantly from the rotation of the rest of the Earth. A rigorous calculation of the so-called gravitational restoring torque acting between the inner core and the rest of the Earth (Szeto & Smylie 1984a) and analytical solutions over a large phase-space (Szeto & Smylie 1988) have shown the 'mechanical' aspects of this proposal to be viable. The motivation behind those studies is an attempt to model the observation that on the Earth's surface the dipole component of the geomagnetic field is tilted with respect to the geographic pole and westward drifting. The hypothesis is essentially as follows: an inclined, ellipsoidal, solid inner core is expected to precess in a mantle-fixed frame. This highly conducting body freezes the dipole within it, giving rise by its tilt and resulting precession to a drifting, inclined dipole field observed on the Earth's surface. A major shortcoming exists, however, in that no consideration was given to the effects on the motions of the inner core due to the presence of dynamo action in the surrounding fluid core.

The magnetic diffusion time scale of a dipole magnetic field into (or out of) the inner core is marginally shorter than the dipole's westward drift period inferred from observations, the latter being about 7000 years (Szeto & Smylie 1984b). Thus an inner-core-frozen field gradually loses strength as it undergoes precession unless it is reinforced. In order to examine the feasibility (or otherwise) of the above model one should evaluate the problem of constructing a magnetohydrodynamical (M.H.D) dynamo capable of producing an inclined magnetic field. This is the purpose of this paper.

Model Description and Assumptions

A kinematic dynamo operating in the fluid outer core is sought, under the following conditions:

1. the Earth is divided into three concentric spherical regions, as shown in Figure 1, with outer radii r_1, r_2 and r_3. The inner core is designated Region I, the outer core Region II, the mantle Region III, and outer space Region IV. Material properties are constant within each region, with respective electrical conductivities $\sigma_1, \sigma_2, \sigma_3$ and σ_4. The last quantity vanishes for an insulating atmosphere and outer space.

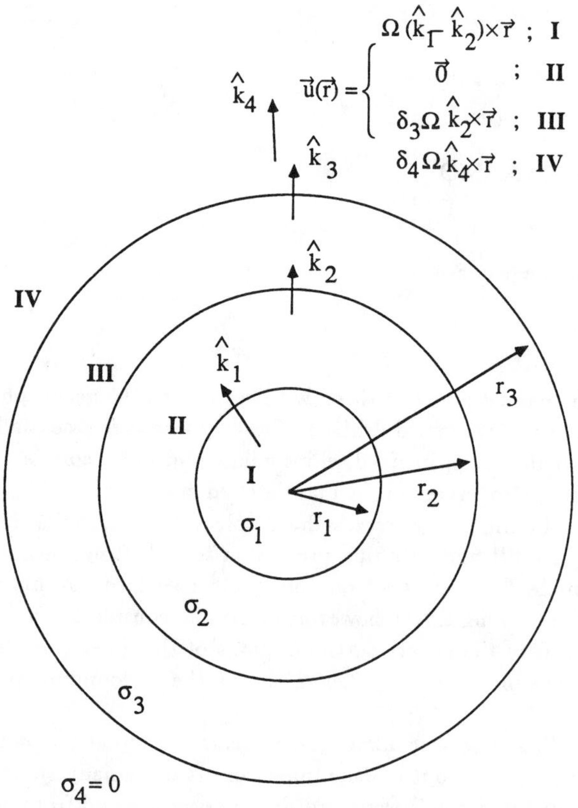

Fig. 1. Adopted Physical Model of the Earth – see text for explanation.

2. the outer core undergoes solid body rotation over appropriate time– and space–scales. i.e. except for small scale turbulence (which must exist in order to allow dynamo action) the fluid exhibits no mean flow relative to a frame uniformly rotating about the $\hat{k}_2$ direction at an average solid body rotation speed Ω. The mantle rotation axis is $\hat{k}_3$, which for simplicity has been taken identical as $\hat{k}_2$, but its rotation rate differs from that of the underlying core by a small factor δ_3. A differential rotation between the mantle and the outer core is thus modelled by a velocity discontinuity across an infinitely thin layer at the core mantle boundary. Another velocity jump occurs at the inner core/outer core interface. The inner core rotates at the same rate as the fluid core as a whole, but about $\hat{k}_1$, substantially tilted relative to $\hat{k}_2$. Both the mantle and inner core exhibit solid body rotation. The Earth as a whole rotates about yet another direction $\hat{k}_4$ at a rate which can be determined by summing spin angular momenta of its constituent shells, using the rotation vectors of the other three regions and their moments of inertia. But since region IV is an insulator, its angular rotation is of no interest to the present dynamo calculation, and will not be further pursued.

3. The only agent for dynamo action in the fluid proper is a spatially constant α-effect. Additional dynamo action occurs at the core/mantle and inner core/outer core boundaries due to differential rotations at these interfaces.

A reference frame is adopted with its origin at the center of the Earth and rotating with angular velocity $\Omega\hat{k}_2$. Thus the velocity field at an arbitrary location with the Earth, according to assumption 2 above, is given by:

$$\vec{u}(\vec{r}) = \begin{cases} \vec{u}_1 = & \Omega(\hat{k}_1 - \hat{k}_2) \times \vec{r} \quad 0 \le r < r_1 \\ & \vec{0} \quad\quad r_1 < r < r_2 \\ \vec{u}_3 = & \delta_3 \Omega \hat{k}_2 \times \vec{r} \quad r_2 < r < r_3 \end{cases} \tag{1}$$

Szeto & Smylie (1984a) made the observation that the dipole westward drift rate is substantially lower than those of the nondipole components. Consistent with that, it will be assumed here that the dipole drift period is 7000 years, while all nondipole components drift westwards with a common period of 2000 years. Though it is well known that the various components move at different rates, adoption of a common rate allows a mathematically simple description of the mantle/core differential rotation. Indeed, the mantle must therefore drift eastward relative to the outer core–fixed reference frame with a 2000 year period, yielding a small positive value of δ_3 equal to 1day/2000 years, or approximately 1.4×10^{-6}.

The remaining quantity to be specified in this kinematic dynamo problem is $\hat{k}_1$. It can be described as a unit vector inclined to the reference direction $\hat{k}_2$ by an angle θ_1, and precessing eastward at the rate ω_1, which numerically equals (1 day/2000 years - 1 day/7000 years), so that the inner core (together with its frozen dipole) is seen to drift westward relative to the mantle with a 7000 year period. In order to produce a surface field inclined at 11° to $\hat{k}_3$, θ_1 should exceed 11° by an amount depending on the attenuation by intervening material between the inner and the surface. As a initial estimate, however, θ_1 is taken simply as 11°.

Induction Equation and Boundary Conditions

The general form of the induction equation governing the magnetic field $\vec{B}$ allowing for an α–effect is

$$\nabla \times \left[\frac{1}{\sigma} \nabla \times (\frac{\vec{B}}{\mu}) - \vec{u} \times \vec{B} - \alpha \vec{B} \right] + \frac{\partial \vec{B}}{\partial t} = \vec{0} \tag{2}$$

Taking magnetic permeability μ and electrical conductivity σ spatially uniform within the four regions, noting that α vanishes everywhere except in region II, and adopting $\mu = \mu_0$ (which applies where temperatures exceed the Curie point), one obtains

Region IV:

$$\nabla \times \nabla \times \vec{B} = \vec{0} \tag{3}$$

Region III:

$$\frac{1}{\mu_0\sigma_3}\nabla \times \nabla \times \vec{B} + \frac{\partial \vec{B}}{\partial t} = \nabla \times (\vec{u}_3 \times \vec{B}) \tag{4}$$

Region II:

$$\frac{1}{\mu_0\sigma_2}\nabla \times \nabla \times \vec{B} + \frac{\partial \vec{B}}{\partial t} = \nabla \times (\alpha \vec{B}) \tag{5}$$

Region I:

$$\frac{1}{\mu_0\sigma_1}\nabla \times \nabla \times \vec{B} + \frac{\partial \vec{B}}{\partial t} = \nabla \times (\vec{u}_1 \times \vec{B}) \tag{6}$$

Since $\vec{B}$ is solenoidal, a toroidal/poloidal decomposition is appropriate. One may write

$$\begin{aligned}\vec{B} = \sum_{n=1}^{\infty}\sum_{m=-n}^{n} &\left\{\nabla \times [\vec{r}T_n^m(r)\mathrm{P}_n^m(\cos\theta)e^{im\phi}]\right. \\ &\left.+\nabla \times \nabla \times [\vec{r}S_n^m(r)\mathrm{P}_n^m(\cos\theta)e^{im\phi}]\right\} e^{im\omega t}\end{aligned} \tag{7}$$

where $T_n^m(r)$ and $S_n^m(r)$ are scalar functions giving rise respectively to toroidal and poloidal magnetic field components. The assumed time dependence corresponds to a field precessing eastwards about the reference $\vec{k}_2$ direction at angular rate $-\omega$.

The assumed forms of $\vec{u}_1$ and $\vec{u}_3$ according to the previous section make them also readily expressible in spherical harmonics, for they are toroidal fields of particularly simple geometry:

$$\vec{u}_3 = \nabla \times [\vec{r}t_1^0(r)\mathrm{P}_1^0(\cos\theta)] \tag{8}$$

where

$$t_1^0(r) = \delta_3\Omega r; r_2 < r < r_3 \tag{9}$$

while

$$\begin{aligned}\vec{u}_1 = \;&\nabla \times [\vec{r}t_1^0(r)\mathrm{P}_1^0(\cos\theta)] \\ &+\nabla \times [\vec{r}t_1^1(r)\mathrm{P}_1^1(\cos\theta)e^{i\phi}]e^{i\omega_1 t} \\ &+\nabla \times [\vec{r}t_1^{-1}(r)\mathrm{P}_1^{-1}(\cos\theta)e^{i\phi}]e^{-i\omega_1 t}\end{aligned} \tag{10}$$

where

$$\left.\begin{aligned}t_1^{-1}(r) &= \Omega\sin\theta_1 r \\ t_1^0(r) &= \Omega(\cos\theta_1 - 1)r \\ t_1^1(r) &= -\tfrac{1}{2}\Omega\sin\theta_1 r\end{aligned}\right\}; 0 \le r < r_1 \tag{11}$$

It can be easily seen that if the magnetic field $\vec{B}$ is to precess westward at a rate ω, the vector $\vec{k}_1$ must precess westward about the reference direction at precisely the same rate. Hence one requires that ω_1 equal ω.

The next step is to extract spherical harmonic components from the terms $\nabla \times \nabla \times \vec{B}$, $\frac{\partial \vec{B}}{\partial t}$, $\nabla \times (\vec{u}_1 \times \vec{B})$, $\nabla \times (\vec{u}_3 \times \vec{B})$ and $\nabla \times (\alpha\vec{B})$. As all of these are divergence-free, a toroidal/poloidal decomposition again suggests itself. Thus the vector induction equation in each of the four regions can be expected to give rise to only two scalar equations (involving T_n^m, S_n^m, t_n^m), the third equation being accounted for by vanishing divergences. The assumption of a constant α-effect enables great simplification for two reasons. First $\nabla \times (\alpha\vec{B})$ is simply $\alpha\nabla \times \vec{B}$, and the rules for extracting toroidal and poloidal components out of this expression are straight forward. Second, as will be shown later, the resulting equations admit analytical solutions in the dynamo region in terms of spherical Bessel functions. As these are precisely the same analytic functions found in the other regions, boundary conditions can be written in a simple algebraic form.

Following standard 'selection rules' (e.g. Smylie 1965) one arrives at the following scalar version of the induction equation:

Region IV:

$$D_n\left\{\begin{array}{c} S_n^m \\ T_n^m \end{array}\right\} = 0 \tag{12}$$

Region III:

$$\frac{1}{\mu_0\sigma_3}D_n\left\{\begin{array}{c} S_n^m \\ T_n^m \end{array}\right\} + im\omega\left\{\begin{array}{c} S_n^m \\ T_n^m \end{array}\right\} = -im\delta_3\Omega\left\{\begin{array}{c} S_n^m \\ T_n^m \end{array}\right\} \tag{13}$$

Region II:

$$\frac{1}{\mu_0\sigma_2}D_n\left\{\begin{array}{c} S_n^m \\ T_n^m \end{array}\right\} + im\omega\left\{\begin{array}{c} S_n^m \\ T_n^m \end{array}\right\} = \alpha\left\{\begin{array}{c} T_n^m \\ D_nS_n^m \end{array}\right\} \tag{14}$$

Region I:

$$\begin{aligned}\frac{1}{\mu_0\sigma_1}\;D_n\left\{\begin{array}{c} S_n^m \\ T_n^m \end{array}\right\} + im\omega\left\{\begin{array}{c} S_n^m \\ T_n^m \end{array}\right\} = \\ -im\Omega(\cos\theta_1 - 1)\left\{\begin{array}{c} S_n^m \\ T_n^m \end{array}\right\} \\ -\frac{i\Omega\sin\theta_1}{2}\left[\begin{array}{c}(n-m)(n+m+1)\left\{\begin{array}{c} S_n^{m+1} \\ T_n^{m+1} \end{array}\right\} \\ +\left\{\begin{array}{c} S_n^{m-1} \\ T_n^{m-1} \end{array}\right\}\end{array}\right]\end{aligned} \tag{15}$$

where

$$D_n = -\left[\frac{d^2}{dr^2} + \frac{2d}{rdr} - \frac{n(n+1)}{r^2}\right] \tag{16}$$

To completely specified the problem, boundary conditions are required. At each of the three material interfaces shown in Figure 1 the magnetic field and the tangential electric field must be continuous. These are more usefully written in terms of continuity in S_n^m, T_n^m, $\frac{dS_n^m}{dr}$ and the tangential part of $\frac{1}{\mu\sigma}(\nabla \times \vec{B}) - \vec{u} \times \vec{B}$. The last two–dimensional vector condition turns out to contribute only one independent scalar condition, namely continuity of the spheroidal transverse component of the electric field. As already noted by Smylie (1965), continuity of the torsional component of the electric field is redundant provided that the field is periodic in time. In addition to these four conditions, $\vec{B}$ must be finite at the center of the Earth and must vanish at infinity.

Solutions

Region IV:

The appropriate induction equation and the boundary condition at infinity yields

$$S_n^m(r) = \frac{(\mathrm{D}_4)_n^m}{r^{n+1}} \tag{17}$$

$$T_n^m(r) = 0 \tag{18}$$

where $(\mathrm{D}_4)_n^m$ is an arbitrary constant which may, in principle, be fitted to magnetic field data at the Earth's surface.

Region III:

Rewriting the induction equation in this region,

$$\left[D_n - k_3^2\right]\left\{ \begin{array}{c} S_n^m \\ T_n^m \end{array} \right\} = 0 \tag{19}$$

where

$$k_3 = \sqrt{\mu_0\sigma_3 m(\omega + \delta_3\Omega)}\frac{-1+i}{\sqrt{2}} \tag{20}$$

Solutions are as follows:

$m \neq 0$:

$$S_n^m(r) = (C_3)_n^m j_n(k_3 r) + (\mathrm{D}_3)_n^m y_n(k_3 r) \tag{21}$$

$$T_n^m(r) = (\mathrm{A}_3)_n^m j_n(k_3 r) + (\mathrm{B}_3)_n^m y_n(k_3 r) \tag{22}$$

$m = 0$:

$$S_n^0(r) = (\mathrm{C}_3)_n^0 r^n + (\mathrm{D}_3)_n^0 r^{-(n+1)} \tag{23}$$

$$T_n^0(r) = (\mathrm{A}_3)_n^0 r^n + (\mathrm{B}_3)_n^0 r^{-(n+1)} \tag{24}$$

where j_n and y_n are respectively spherical Bessel functions of the first and second kind of order n, and $(\mathrm{A}_3)_n^m, (\mathrm{B}_3)_n^m, (\mathrm{C}_3)_n^m, (\mathrm{D}_3)_n^m$ are constants that may be related to $(\mathrm{D}_4)_n^m$ via boundary conditions at the Earth's surface.

Region II:

The induction equation may be rewritten as

$$(D_n - k_+^2)(D_n - k_-^2)\left\{ \begin{array}{c} S_n^m \\ T_n^m \end{array} \right\} = 0 \tag{25}$$

where

$$k_\pm = \frac{\mu_0\sigma_2\alpha}{2}\left[1 \pm \sqrt{1 - i\frac{4m\omega}{\mu_0\sigma_2\alpha^2}}\right] \tag{26}$$

Solutions are

$m \neq 0$:

$$\begin{aligned} S_n^m = \; & (\mathrm{A}_2)_n^m j_n(k_+ r) + (\mathrm{B}_2)_n^m j_n(k_- r) \\ & +(\mathrm{C}_2)_n^m y_n(k_+ r) + (\mathrm{D}_2)_n^m y_n(k_- r) \end{aligned} \tag{27}$$

$$\begin{aligned} T_n^m = \; & k_+(\mathrm{A}_2)_n^m j_n(k_+ r) + k_-(\mathrm{B}_2)_n^m j_n(k_- r) \\ & +k_+(\mathrm{C}_2)_n^m y_n(k_+ r) + k_-(\mathrm{D}_2)_n^m y_n(k_- r) \end{aligned} \tag{28}$$

$m = 0$:

$$S_n^0 = (\mathrm{A}_2)_n^0 j_n(k_2 r) + (\mathrm{B}_2)_n^0 r^n + (\mathrm{C}_2)_n^0 y_n(k_2 r) + (\mathrm{D}_2)_n^0 r^{-(n+1)} \tag{29}$$

$$T_n^0 = k_2(\mathrm{A}_2)_n^0 j_n(k_2 r) + k_2(\mathrm{C}_2)_n^0 y_n(k_2 r) \tag{30}$$

where only one value of $k_\pm$ remains in this degenerate case:

$$k_2 = \mu_0\sigma_2\alpha \tag{31}$$

Again $(\mathrm{A}_2)_n^m, (\mathrm{B}_2)_n^m, (\mathrm{C}_2)_n^m$ and $(\mathrm{D}_2)_n^m$ are constants related to their counterparts in neighbouring regions via boundary conditions.

Region I:

While not coupled across degree n, the induction equation forms a coupled set in order m as a consequence of non-axisymmetric $t_1^{\pm 1}$ toroidal velocity components in $\vec{u}_1$. Accounting for the regularity condition at the origin, T_n^m and S_n^m are expressible as linear combinations of r^n and spherical Bessel functions of the first kind:

$m \neq 0$:

$$S_n^m(r) = (\mathrm{D}_1)_n^m r^n + \sum_{j=1}^{n} (\mathrm{C}_{1,j})_n^m j_n(k_j r) \tag{32}$$

$$T_n^m(r) = (\mathrm{B}_1)_n^m r^n + \sum_{j=1}^{n} (\mathrm{A}_{1,j})_n^m j_n(k_j r) \tag{33}$$

$m = 0$:

$$S_n^0(r) = (\mathrm{D}_1)_n^0 r^n \tag{34}$$

$$T_n^0(r) = (\mathrm{B}_1)_n^0 r^n \tag{35}$$

where k_j are eigenvalues of the coupled system of n differential equations applicable to a specific value of n, and $(A_{1,j})_n^m, (C_{1,j})_n^m$ are coefficients of eigenvectors of arbitrary lengths. It can be shown that $(B_1)_n^m$ and $(D_1)_n^m$ are dependent over order m, giving rise to only one arbitrary constant within each set. Thus the poloidal and toroidal systems each contains $n+1$ constants to be determined from boundary conditions. A similar coupled system to this was first discussed by Rochester (1968), but where $2n+1$ constants were needed because the internal boundary was a sphere rather than a point.

Eigenvalue Problem in α

Applying the condition of continuity in S_n^m and T_n^m at the Earth's surface yields the following relationships:

$m \neq 0$:

$$(B_3)_n^m = (A_3)_n^m \left[-\frac{j_n(k_3 r_3)}{y_n(k_3 r_3)}\right] \tag{36}$$

$$\left\{ \begin{array}{c} (C_3)_n^m \\ (D_3)_n^m \end{array} \right\} = \frac{k_3^2 (D_4)_n^m}{r_3^{n-1}} \left\{ \begin{array}{c} \frac{n+1}{k_3 r_3} y_n(k_3 r_3) + y_n'(k_3 r_3) \\ -\frac{n+1}{k_3 r_3} j_n(k_3 r_3) - j_n'(k_3 r_3) \end{array} \right\} \tag{37}$$

$m = 0$:

$$(B_3)_n^0 = -(A_3)_n^0 r_3^{2n+1} \tag{38}$$

$$(C_3)_n^0 = 0 \tag{39}$$

$$(D_3)_n^0 = (D_4)_n^0 \tag{40}$$

Evidently for a prescribed pair of values of (n, m) only two adjustable parameters A_3 and D_4 are required to describe the solutions for S_n^m and T_n^m.

Boundary conditions are next applied at the core–mantle boundary. Since S_n^m is specified by $(D_4)_n^m$ on the mantle side and by four constants $(A_2)_n^m, (B_2)_n^m, (C_2)_n^m, (D_2)_n^m$ on the core side, continuity conditions for S_n^m and its radial derivative may be rewritten in the form

$$[f_1(A_2, B_2, C_2, D_2, D_4)]_n^m = 0 \tag{41}$$

$$[f_2(A_2, B_2, C_2, D_2, D_4)]_n^m = 0 \tag{42}$$

which may be used to eliminate D_4 in generating a 'consistency' condition involving only the four core coefficients of the form:

$$[g_1(A_2, B_2, C_2, D_2)]_n^m = 0 \tag{43}$$

which is a linear homogeneous condition, i.e. a linear combination of these four coefficients which must vanish.

While the continuity of T_n^m also yields a simple equation of the form

$$[f_3(A_2, B_2, C_2, D_2, A_3)]_n^m = 0 \tag{44}$$

the continuity of the tangential electric field unfortunately reads

$$\left[\frac{1}{\mu_0 \sigma_2} \nabla \times \vec{B} - \alpha \vec{B}\right]_{r_2^-} = \left[\frac{1}{\mu_0 \sigma_3} \nabla \times \vec{B} - \vec{u}_3 \times \vec{B}\right]_{r_2^+} \tag{45}$$

whose spheroidal transverse part reveals coupling of degree n terms with degree $n \pm 1$ terms (Smylie & Mansinha 1971, Rochester 1968):

$$\left[\frac{1}{\mu_0 \sigma_2} \frac{1}{r} \frac{d}{dr}(r T_n^m) - \frac{1}{r} \alpha \frac{d}{dr}(r S_n^m)\right]_{r_2^-} = \left[\frac{1}{\mu_0 \sigma_3} \frac{1}{r} \frac{d}{dr}(r T_n^m) + \delta_3 \Omega r \left\{\frac{n-m}{2n-1}(n-1) S_{n-1}^m - \frac{n+m+1}{2n+3}(n+2) S_{n+1}^m\right\}\right]_{r_2^+} \tag{46}$$

This leads to a condition of the type

$$f_4\left((D_4)_{n-1}^m, (A_2, B_2, C_2, D_2, A_3, D_4)_n^m, (D_4)_{n+1}^m\right) = 0 \tag{47}$$

Eliminating A_3 and D_4 with the aid of f_3 and f_2, one obtains a second consistency condition:

$$g_2 \left(\begin{array}{c} (A_2, B_2, C_2, D_2)_{n-1}^m, (A_2, B_2, C_2, D_2)_n^m, \\ (A_2, B_2, C_2, D_2)_{n+1}^m \end{array} \right) = 0 \tag{48}$$

Finally boundary conditions are applied at the inner core/outer core boundary. Continuity in S and $\frac{d}{dr}S$ give rise to conditions of the form

$$[f_5(A_2, B_2, C_2, D_2, D_1, C_{1,1}, \ldots, C_{1,j}, \ldots, C_{1,n})]_n^m = 0 \tag{49}$$

$$[f_6(A_2, B_2, C_2, D_2, D_1, C_{1,1}, \ldots, C_{1,j}, \ldots, C_{1,n})]_n^m = 0 \tag{50}$$

For a given degree n, there are $n+1$ equations of the forms f_5 and f_6. These two sets of equations produce a third homogeneous consistency equation

$$g_3 \left(\begin{array}{c} (A_2, B_2, C_2, D_2)_n^0, \ldots (A_2, B_2, C_2, D_2)_n^m, \\ \ldots (A_2, B_2, C_2, D_2)_n^n \end{array} \right) = 0 \tag{51}$$

Like the complication that arises in g_2, the final consistency condition involves core coefficients of degree and order other than n and m. Indeed, due to coupling in $\vec{u}_1 \times \vec{B}$ in the electric field, $n \pm 1, m \pm 1$ terms are found in the continuity of its spheroidal transverse component, which reads

$$\begin{aligned} \left[\frac{1}{\mu_0 \sigma_2} \frac{1}{r} \frac{d}{dr}(r T_n^m) - \frac{1}{r} \alpha \frac{d}{dr}(r S_n^m)\right]_{r_1^+} = \Bigg[& \frac{1}{\mu_0 \sigma_1} \frac{1}{r} \frac{d}{dr}(r T_n^m) \\ & + \Omega r \sin\theta_1 \left\{ \begin{array}{c} \frac{(n-m-1)(n-m)}{2(2n-1)}(n-1) S_{n-1}^{m+1} \\ + \frac{(n+m+1)(n+m+2)}{2(2n+3)}(n+2) S_{n+1}^{m+1} \end{array} \right\} \\ & + \Omega r (\cos\theta_1 - 1) \left\{ \begin{array}{c} \frac{n-m}{2n-1}(n-1) S_{n-1}^m \\ - \frac{n+m+1}{2n+3}(n+2) S_{n+1}^m \end{array} \right\} \\ & + \Omega r \sin\theta_1 \left\{ \begin{array}{c} -\frac{1}{2(2n-1)}(n-1) S_{n-1}^{m-1} \\ - \frac{1}{2(2n+3)}(n+2) S_{n+1}^{m-1} \end{array} \right\} \Bigg]_{r_1^-} \end{aligned} \tag{52}$$

This and the continuity of T_n^m together yield

$$g_4\begin{pmatrix}(A_2,B_2,C_2,D_2)^0_{n-1},\ldots,(A_2,B_2,C_2,D_2)^{n-1}_{n-1},\\(A_2,B_2,C_2,D_2)^0_n,\ldots,(A_2,B_2,C_2,D_2)^n_n,\\(A_2,B_2,C_2,D_2)^0_{n+1},\ldots,(A_2,B_2,C_2,D_2)^{n+1}_{n+1}\end{pmatrix}=0 \tag{53}$$

Implementation of the four conditions (43), (48), (51) and (53) is depicted in Figure 2 in matrix form. Each cross represents a non-zero entry whose value is dependent on α. Values of α for which this (infinite) matrix vanishes are eigenvalues of this dynamo model when the four boundary conditions are satisfied for all degree n and order m harmonic components. Groups of four rows represent boundary conditions for appropriate pairs (n,m); the first in each group corresponding to g_1 vanishing, and so on with the other three. The groups are arranged in the following order of (n,m): $(1,0),(1,1),(2,0)\ldots(2,2),\ldots(n,0)\ldots(n,m)\ldots(n,n)$. Eigenvectors of this matrix yield A_2, B_2, C_2, D_2 to within a constant amplitude factor, also grouped in the same order.

Results

An α^2 dynamo operating in a unit sphere, where α is spatially constant, has previously studied (Krause & Steenbeck 1967, Roberts 1971). The critical magnetic Reynolds number, R_{crit}, defined as $\mu_0\sigma r\alpha$, is the first zero of j_1, approximately 4.49. If α is allowed to vary with colatitude as in the simple model $\alpha=\alpha_1^0\cos\theta$ then there are two 'chains' of spherical harmonics with similar magnetic Reynolds numbers: ~ 7.6 for the dipole chain, and ~ 7.8 for the quadrupole chain (Steenbeck & Krause 1966). These larger values over the case of constant α can perhaps be understood in terms of the fact that a $\cos\theta$ dependence gives rise to less effective dynamo action over the entire sphere, especially near the equator. A variety of spatial dependence of α in a unit sphere dynamo have also been considered (Roberts 1972, Rädler 1986, Rädler & Bräuer 1987).

St. Pierre (1987) considered dynamo action in a spherical shell of unit thickness, assuming $\alpha=\alpha_1^0\cos\theta$ and introducing an adjustable differential rotation at the inner spherical boundary. Turning off this differential rotation gives R_{crit} of ~ 5.19 and ~ 5.27 respectively for the dipole and quadrupole chains. These values are lower than corresponding ones over the spherical dynamo because scaling the shell thickness to unity gives a larger volume in which dynamo action takes place. Corresponding numbers for a exaggeratedly fast differential rotation are 3.2 and 4.3 which are lower still, presumably due to enhanced dynamo action arising from the $\omega-$effect.

In the present study the dynamo resides in a shell where α is constant and whose dimensionless inner and outer radius are ~ 0.19 and ~ 0.55. Disregarding $\omega-$effect from differential rotation for the moment, one may scale the unit sphere result to serve as a lower bound on R_{crit}, since removal of an dynamo-active region in the center represents weakened dynamo action. As the strength of dynamo action is proportional to the square of the dimension of the dynamo, this yields a lower bound of ~ 14. On the other hand introduction of non-vanishing differential rotations at both boundaries of the dynamo should lower the critical value somewhat. A value between 10 and 30 may be expected.

Discussion

While calculations are currently under way, no firm numerical values can yet be reported. However, a few tentative trends have been found:

- Complex nature of α: a non-zero imaginary part in R_{crit} is persistent in all the calculations so far. If this is is not an artifact of the analysis, the $\alpha-$effect produces an electromotive force which lags or leads the magnetic field.
- Dependence on θ_1: beginning with a small tilt of the inner core rotation axis, R_{crit} seems to decrease systematically as θ_1 is increased. This may be understood in terms of increased ω dynamo action resulting from increased differential rotation, which reduces the demand on the $\alpha-$effect to maintain the magnetic field against decay.

A constant α has been adopted in this paper for lack of knowledge that suggests an alternative. But if one were to introduce a spatially varying $\alpha-$effect, one would expand α in terms of a scalar spherical harmonic series and derive an alternative set of induction equations for region II. The term immediately after the constant term in such a series corresponds to an α-effect proportional to the cosine of colatitude. A version of the resulting induction equation in Region II for the case of zero differential rotations at the boundaries can be found in Smylie et al. (1984). However, it is evident that the induction equation in the dynamo region for a spatially varying α will not in general admit simple closed-form solutions for S_n^m and T_n^m. Some kind of finite element numerical scheme must then be employed.

Future Work

The first task will be experimentation with adjustable parameters θ_1, δ_3 and ω in order to understand their in-

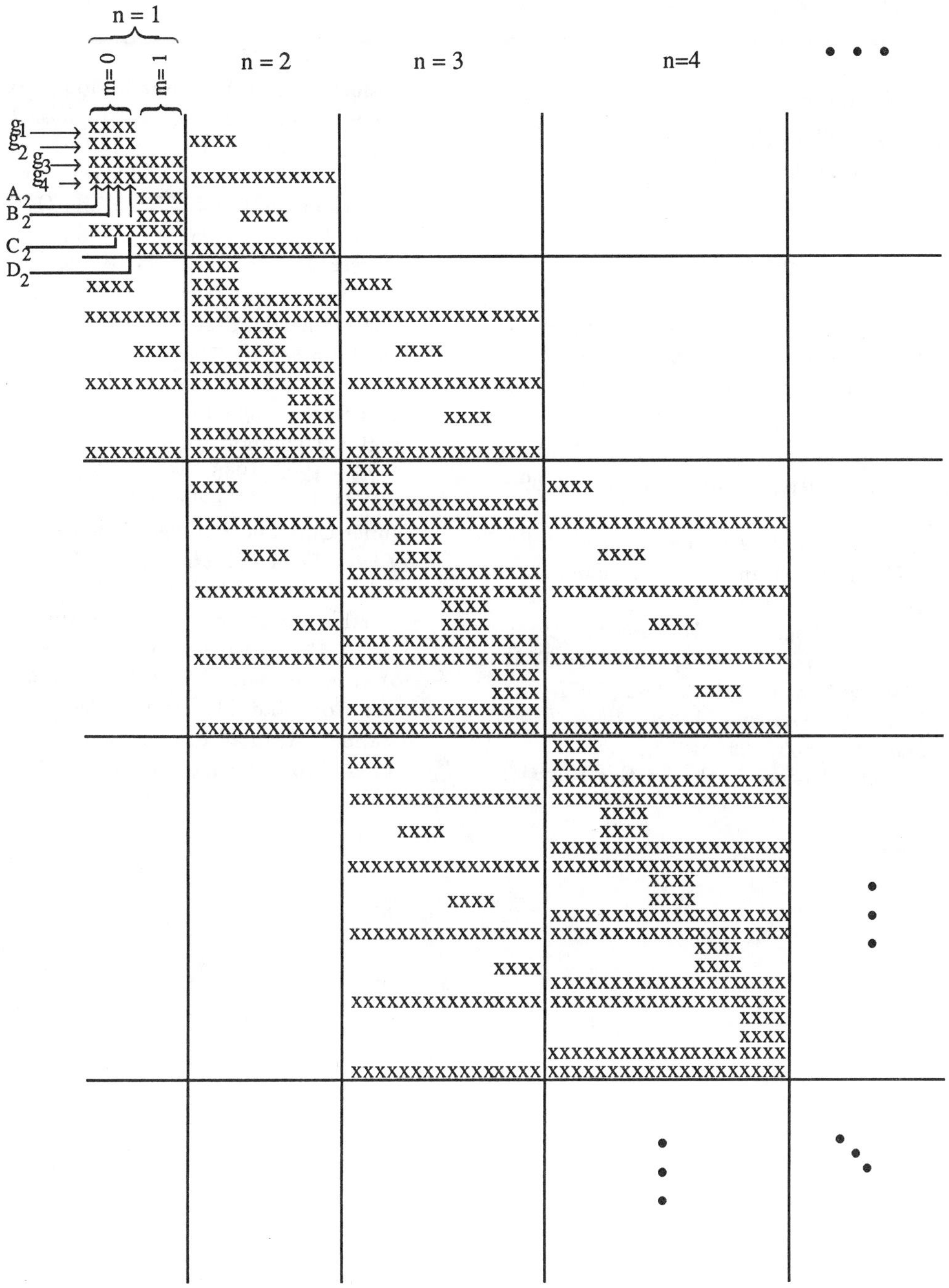

Fig. 2. Schematic of Eigenvalue Matrix – see text for explanation.

fluence on the inclined dynamo model. Solving for the relative magnitudes of the core coefficients A_2, B_2, C_2, D_2 would allow the importance of differential rotation to be evaluated – whether the toroidal field in the dynamo is comparable to or much stronger than the poloidal field. Given the velocity field, and having calculated the magnetic field, one may estimate the electromagnetic torque at the core/mantle and inner core/outer core interfaces, which may be fed back into a more comprehensive scheme including the dynamical response of the inner core and of the outer core fluid, thus leading towards a dynamically self–consistent MHD dynamo that takes into account the presence of an conducting inner body as well as possible internal differential rotation.

Acknowledgments. The author is grateful to Dr D. E. Smylie and Dr K. Sato for helpful discussions.

References

Krause F. and Steenbeck M., Some Simple Models of Magnetic Field Regeneration by Non-Mirror Symmetric Turbulences, *Z. Naturforsch. 22a,* 671 – 675, 1967.

Rädler K.H., Investigations of Spherical Kinematic Mean – Field Dynamo Models, *Astron. Nachr. 307,* 89 – 113, 1986.

Rädler K.H. and Bräuer H.J., On the Oscillatory Behaviour of Kinematic Mean – Field Dynamo, *Astron. Nachr. 308,* 101 – 109, 1987.

Roberts P.H., Dynamo Theory, in *Mathematical Problems in the Geophysical Sciences,* edited by W.H. Reid, pp. 129 – 206, Amer. Math. Soc., 1971.

Roberts P.H., Kinematic Dynamo Models, *Phil. Trans. R. Soc. Lond.A272,* 663 – 698, 1972.

Rochester M.G., Perturbations in the Earth's Rotation and Geomagnetic Core-Mantle Coupling, *J. Geomag. Geoelectr. 20,* 387 – 402, 1968.

Smylie D.E. and Mansinha L., The Elastic Theory of Dislocations an Real Earth Models and Changes in the Rotation of the Earth, *Geophys. J. R. astr. Soc. 23,* 329 – 354, 1971.

Smylie D.E., Szeto A.M.K. and Rochester M.G., The Dynamics of the Earth's Inner and Outer Cores, *Rep. Prog. Phys. 47,* 855 – 906, 1984.

St. Pierre M., Implications of the Vanishing of the Electromagnetic Torques at the Boundaries of an Axisymmetric, Steady-State Dynamo, *MSc Thesis, York University,* 1987.

Steenbeck M. and Krause F., Erklärung stellarer und planetarer Magnetfelder durch einen turbulenzbedingten Dynamomechanismus *Z. Naturforsch. 21a,* 1285 – 1296, 1966.

Szeto A. M. K. and Smylie D. E., Coupled Motions of the Inner Core and Possible Geomagnetic Implications, *Phys. Earth Planet. Int.36,* 27 – 42, 1984a.

Szeto A. M. K. and Smylie D. E., The Rotation of the Earth's Inner Core, *Phil. Trans. R. Soc. Lond. A313,* 171 – 184, 1984b.

Szeto A. M. K. and Smylie D. E., Motions of the Inner Core and Mantle Coupled via Mutual Gravitation: Regular Precessional Modes, accepted by *Phys. Earth Planet. Int.* in Jan, 1988.

INSTABILITIES OF TOROIDAL MAGNETIC FIELDS

David R. Fearn

Department of Mathematics, University of Glasgow, University Gardens, Glasgow, G12 8QW, UK

Abstract. Why are we interested in the stability of toroidal magnetic fields in rapidly rotating systems? To answer this question we begin by reviewing the geophysical background to the problem and discuss the constraints that magnetic instabilities may impose on the strength and configuration of planetary magnetic fields. Instabilities may also play a role in the geomagnetic secular variation and in reversals. We introduce a simple mathematical model and show that the various instabilities that have been found can be understood simply. The most important fall into one of two classes: ideal or resistive. We discuss the conditions required for instability, the instability timescale, and various characteristic features of each class. In the literature, several other instabilities have been identified. We propose that they may all be understood in the following manner. The effect of rotation on the ideal instability is stabilising. If some effect can be found to counteract the rotational constraint, then instability may arise. Destabilising ingredients are found to include stratification, fluid inertia, viscosity, and an axial magnetic field.

Introduction

Dynamo theory is concerned with the problem of how a magnetic field is maintained against ohmic decay by its interaction with the motion of an electrically conducting fluid. The investigation of the stability of the generated field is an important complementary study. Waves travelling along the field lines may contribute to the observed secular variation [see for example Bloxham and Gubbins,1985]. If such waves become unstable, an excursion or a reversal [see for example Jacobs,1984] may occur. Typically, as we shall see, a magnetic field must satisfy two conditions to become unstable. The geometry of the field must be suitable, and the field strength must be sufficiently large that the energy supplied to the instability is greater than that required to overcome ohmic losses. A comprehensive understanding of magnetic instabilities will permit predictions to be made about what field geometries and strengths are physically reasonable. Such studies are likely to become of increasing importance with the prospect in the near future of attempts to solve the full nonlinear hydromagnetic dynamo problem numerically. A thorough knowledge of the conditions under which magnetic fields become unstable, and the characteristics of such instabilities will be an essential tool for the interpretation of the results from such numerical models.

Mathematical Model

A realistic model of the Earth's core would use a spherical geometry with an inner core, and incorporate both poloidal and toroidal components of the magnetic field. Other important ingredients such as stratification and differential rotation should also be included. An understanding of such a complicated model must be our eventual aim, but it makes sense to begin with a model that is much simpler while retaining the essential physics of the problem. Since field curvature is important, we choose to study the magnetic field

$$\underline{B} = B(s)\, \underline{1}_\varphi \,, \tag{1}$$

where (s,φ,z) are cylindrical polar coordinates and $\underline{1}_\varphi$ is the unit vector in the direction of increasing φ. The field permeates a fluid of density ρ, magnetic diffusivity η and kinematic viscosity ν which is confined between the walls of an infinite cylindrical container of inner radius s_i and outer radius s_o. The whole system rotates with uniform angular velocity

$$\underline{\Omega}_0 = \Omega_0\, \underline{1}_z \,. \tag{2}$$

The walls are rigid and may be either perfect electrical conductors or insulators.

The equations governing linear perturbations $\underline{b}$ to the magnetic field and $\underline{u}$ to the flow are given (for example) in Fearn [1983]. The coefficients in the equations and the boundary conditions contain no explicit dependence on time t, azimuthal angle φ or axial distance z, so a modal expansion

$$\underline{b}(\underline{r},t) = \underline{b}(s) \exp i(m\varphi+nz) \exp (p-i\omega)t, \quad (3)$$

is permitted. The problem then reduces to solving an ordinary differential equation in s for the eigenvalue $p-i\omega$. Detailed numerical calculations have been carried out by Fearn [1983,1984,1985, 1988] which complement analytic results by Acheson [1972,1973,1983] and Roberts and Loper [1979]. We review the results below.

Ideal Instabilities

In a non-diffusive, non-rotating, magnetic system, the characteristic timescale is Ω_M^{-1} where the Alfvén frequency Ω_M is given by

$$\Omega_M^2 \equiv B^2/\mu\rho s_o^2, \quad (4)$$

where B is a typical magnitude of the toroidal field B, and μ is the magnetic permeability of the fluid. In a rapidly rotating system, such that $\Omega_0 \gg \Omega_M$ (such as the Earth), there are two characteristic timescales: Ω_0^{-1} and

$$\tau_S \equiv 2\Omega_0/\Omega_M^2, \quad (5)$$

see for example Fearn et al [1988]. The former (Ω_0^{-1}) is the inertial timescale and is sometimes referred to as the "fast" timescale. In comparison τ_S is "slow", ($\tau_S \gg \Omega_0^{-1}$).

When diffusion is absent, Acheson [1972] has used a local analysis to investigate the stability of (1). (This approximate analytic method is necessary because the problem is too complicated to permit an exact analytical solution.) The method assumes the lengthscale of the instability $\underline{b}$ to be small compared with that of the basic state $\underline{B}$ and ignores the influence of boundaries. Such a method clearly has its limitations; the results are only approximate and, most importantly, it can only identify certain classes of instability. In the method's favour is its simplicity and its ability to provide a good qualitative feel for the conditions required for instability [see (6) below]. If the local analysis predicts instability, then we expect the system to be unstable. If it predicts no instability, then instability, not local in nature, may still be present. It is therefore essential to complement any local analysis with a full numerical solution of the problem, [see Fearn,1983,1984,1985,1988].

Acheson [1972] finds that $\underline{B}$ is locally unstable where

$$\Delta \equiv (s^3/B^2)\, d(B^2/s^2)/ds > m^2 . \quad (6)$$

He calls this instability the "field gradient instability". It is the rotating counterpart of the "kink" instability well known in plasma physics [see for example Bateman,1978; Wesson, 1981]. The condition (6) applies only for non-axisymmetric instabilities ($m \geq 1$). The axisymmetric case ($m = 0$) was considered by Michael [1954] who showed the effect of rotation to be very strongly stabilising. When $\Omega_0 \gg \Omega_M$, the m = 0 mode is stable unless the magnetic field gradient is very large; the local condition for instability is $s\,d(B^2/\mu\rho s^2)/ds > 4\Omega_0^2$, see equation (5.4) of Acheson [1972]. The most unstable mode is therefore m = 1, so instability requires that somewhere in the interval $s_i < s < s_o$, the toroidal field strength B(s) increases with s faster than $s^{3/2}$. We note that this is consistent with a global, necessary condition for instability

$$\Delta > m^2 - 4, \qquad m > 1 ,$$
$$\Delta > -5, \qquad m = 1 , \quad (7)$$

obtained by Acheson [1973].

The local results have been checked by a numerical solution of the governing equations by Fearn [1983] for the case of perfectly conducting boundaries and Fearn [1988] for the case of insulating boundaries. The numerical results are wholly consistent with (6). If a field B(s) is chosen that is locally unstable in some interval of s and elsewhere locally stable, the numerical results show a tendency for the instability to be concentrated approximately in the locally unstable region [see Figures 3 and 4 of Fearn, 1983].

So far, our discussion of the ideal instability has neglected magnetic diffusion. This is justified in the limit of vanishing diffusivity ($\eta \to 0$) since then it plays no role, but if diffusion is sufficiently strong, it can act to damp out the ideal instability even when (6) is satisfied. An extension of (6) to include the effects of diffusion is given in Acheson [1983], [see also Fearn,1983]. Diffusion becomes important when the diffusive timescale

$$\tau_\eta = s_o^2/\eta , \quad (8)$$

is comparable with the instability timescale; i.e. we expect that the instability will become damped when $\tau_\eta \leq O(\tau_S)$. Equivalently, we expect to find instability only when both (6) and

$$\Lambda \equiv \tau_\eta/\tau_S > \Lambda_c \gg O(1), \quad (9)$$

are satisfied. The dimensionless parameter Λ is known as the Elsasser number. It is more familiar

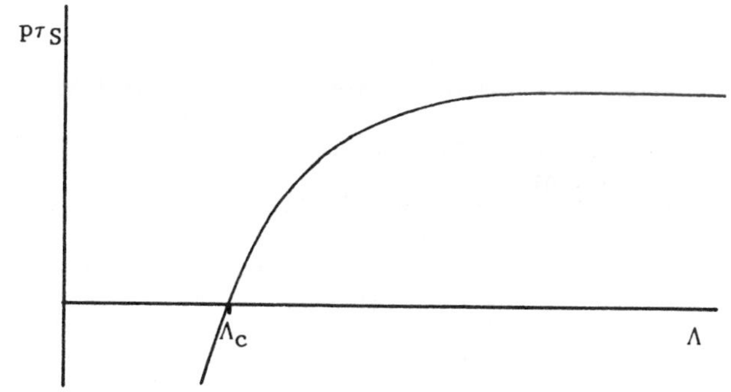

Fig. 1. A sketch of the typical behaviour of the growth rate p of the ideal mode versus the Elsasser number Λ.

as a measure of the relative strength of the Lorentz force compared with the Coriolis force. Here, it provides an inverse measure of the strength of diffusion. A sketch of the behaviour of the growth rate p versus Λ is shown in Figure 1, with $p\tau_S \to$ constant as $\Lambda \to \infty$.

Two further characteristic features of the ideal mode should be mentioned. The first is the direction of propagation. From the local analysis, the frequency is given by

$$\omega = -mB^2/\mu\rho\Omega_0 s^2, \quad (10)$$

[see Fearn,1983, or Acheson,1983], so the instability takes the form of a westwardly propagating wave with phase speed $\omega/m = O(\tau_S{}^{-1})$. The second characteristic feature is the behaviour of the instability when a toroidal flow

$$\underline{U} = R_m(B^2/2\Omega_0\mu\rho s)\ \underline{1}_\varphi , \quad (11)$$

relative to the uniform rotation (2) is introduced into the basic state. In the definition (6) of Δ, B^2/s^2 is then replaced by $B^2/s^2 - 2\Omega_0\mu\rho U/s = (1-R_m)(B^2/s^2)$, (where $\underline{U} = U\underline{1}_\varphi$), so as the parameter R_m is increased from zero, $\Delta \to 0$ as $R_m \to 1$. Hence for any field B(s) that satisfies (6) somewhere, we expect to find that instability disappears as the strength of the toroidal flow (11) is increased. If we focus on the critical Elsasser number, we find $\Lambda_c \to \infty$ as $R_m \to O(1)$. Two examples are illustrated in Figure 2 of Fearn [1983].

Resistive Instabilities

The role of ohmic diffusion can be more subtle than the simple damping effect it has on the ideal instability. Diffusion permits the breaking and reconnection of magnetic field lines and the motion of those field lines relative to the fluid. (In ideal magnetohydrodynamics the field lines are 'frozen' into the fluid and cannot be broken.) In a non-rotating fluid, resistive instability is associated with "resonant surfaces" where $\underline{k}.\underline{B} = 0$, where $\underline{k}$ is the wavenumber of the instability. This is also true in the case of a rapidly rotating system, see Fearn [1984,1985]. In the case of field (1), the condition for instability reduces to B = 0.

When diffusion is weak ($\Lambda \gg 1$), the instability is concentrated in a narrow (critical) layer of width $O(\Lambda^{-1/3})$ about the point where B = 0, but when diffusion is strong, the instability is no longer localised near a zero of B. As for the ideal instability, strong diffusion acts to stabilise the resistive instability. In the limit of vanishing diffusivity, the growth rate $p = O(\tau_S{}^{-2/3}\tau_\eta{}^{-1/3})$ [see Fearn,1984], so p behaves as shown in Figure 2, with $p\tau_S \to 0$ as $\Lambda \to \infty$. The mode may propagate eastward or westward and is not significantly affected by the differential rotation (11).

Mixed Modes

So far, we have treated the ideal and resistive modes as distinct. This is not always so. A field that may be typical of the the Earth's field is one that is zero at the inner boundary, increases in strength with radius s, then decreases to zero at the outer boundary. Such a field may satisfy the conditions for both ideal and resistive instability and examples have been studied by Fearn [1983,1988]. Particularly for the case of insulating boundaries [Fearn, 1988], he finds modes of instability (for m = 1) that show a mixture of the characteristics of the ideal and resistive modes. For example, one mode behaves like an ideal mode when the differential rotation (11) is introduced, but may propagate in either eastward or westward directions. Another mode may be largely unaffected by the differential rotation but otherwise behaves like an ideal mode. See Fearn [1988] for details.

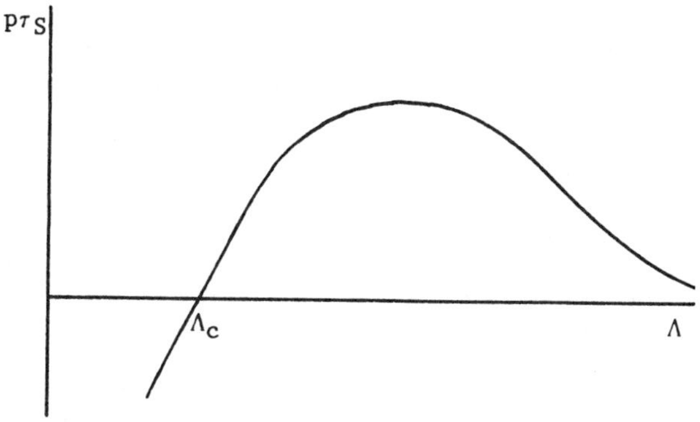

Fig. 2. Typical behaviour of the growth rate p of the resistive instability versus the Elsasser number Λ.

Rotational Constraint and Instability

In a non-rotating system, the local condition for ideal instability is given by

$$\Delta > m^2 - 4 , \qquad (12)$$

[see Fearn, 1985], corresponding to instability for a field B(s) that, somewhere, increases with s faster than $s^{-1/2}$. Comparing (12) with (6), we see that rotation has a strong stabilising influence on the ideal mode. In the case of thermal convection we are familiar with the stabilising influence of rotation and how this can be counteracted by a magnetic field, see for example Eltayeb and Roberts [1970]. It may be that some effect can counteract the rotational constraint in the case of magnetic instability. We propose that various (rather esoteric) instabilities that have been found can be explained in this way.

There are two classes of study in the literature. Those for the field $B \propto s$, for which $\Delta = 0$, and those for arbitrary B(s). For the former we expect neither ideal nor resistive instability, but perhaps instability of the rotational-constraint-breaking category for m = 1 only. Several instabilities have been found. Roberts and Loper [1979] found instability when $-1 < \omega\tau_S < 0$ for the field $B \propto s$ for m = 1 only. This "exceptional" mode requires the presence of non-perfectly-conducting walls. Fluid inertia and magnetic diffusion are also essential ingredients. This mode has been further investigated by Fearn [1988] who showed that fluid viscosity is an alternative to inertia and magnetic diffusion. Fluid viscosity and inertia are very weak effects in the Earth (on the timescale τ_S) and are normally neglected in the so-called magnetostrophic approximation [see for example Fearn et al,1988]. Fearn [1988] has shown that the field strengths (as measured by the critical Elsasser number Λ_c) required for instability are probably much too large for these "exceptional" instabilities to be relevant to the Earth. Roberts and Loper [1979] also proposed that stratification (either top or bottom heavy) could be a destabilising agent. This was investigated further by Soward [1979], Fearn [1979,1983] and Acheson [1980,1983]. Acheson [1983] uses a local analysis to show that there is instability when $\Delta > m^2 - 4$. This suggests that stratification has completely counteracted the rotational constraint. The studies by Soward [1979] and Fearn [1979] show instability for m = 1 in the case $\Delta = 0$. The numerical results of Fearn [1983] confirm Acheson's [1983] local results. Finally, Fearn [1985] investigated the stability of the field $\underline{B} = B(s)\underline{1}_\varphi + B_z\underline{1}_z$. The addition of the axial field was found to be destabilising. In particular, Fearn [1985] found instability for fields satisfying (12) but not (6).

Summary and Concluding Remarks

From our cylindrical model we have built up an understanding of the conditions required for the instability of a toroidal magnetic field. For ideal instability, the field $\underline{B} = B(s)\underline{1}_\varphi$ must increase in strength with cylindrical radius s faster than $s^{3/2}$, somewhere in the core. Resistive instability requires that B(s) vanish somewhere in the core. In both cases, an additional requirement for instability is that the Elsasser number Λ exceed some critical value $\Lambda_c > O(1)$. (Using the molecular value for the magnetic diffusivity, an Elsasser number of order unity corresponds to a field strength of order 2×10^{-3} T.) When B(s) does not satisfy the condition for ideal instability, instability may still be possible if B(s) increases outward faster than $s^{-1/2}$ somewhere and some effect such as stratification or an axial field is present to counteract the constraint of rotation.

References

Acheson, D.J., On the hydromagnetic stability of a rotating fluid annulus, J. Fluid Mech. 52, 529-541, 1972.

Acheson, D.J., Hydromagnetic wavelike instabilities in a rapidly rotating stratified fluid, J. Fluid Mech. 61, 609-624, 1973.

Acheson, D.J., 'Stable' density stratification as a catalyst for instability, J. Fluid Mech. 96, 723-733, 1980.

Acheson, D.J., Local analysis of thermal and magnetic instabilities in a rapidly rotating fluid, Geophys. Astrophys. Fluid Dynam. 27, 123-136, 1983.

Bateman, G., MHD Instabilities, MIT Press, Cambridge, Massachusetts, 1978.

Bloxham, J. and Gubbins, D., The secular variation of Earth's magnetic field, Nature 317, 771-781, 1985.

Eltayeb, I.A. and Roberts, P.H., On the hydromagnetics of rotating fluids, Astrophys. J. 162, 699-701, 1970.

Fearn, D.R., Thermal and magnetic instabilities in a rapidly rotating fluid sphere, Geophys. Astrophys. Fluid Dynam. 14, 103-126, 1979.

Fearn, D.R., Hydromagnetic waves in a differentially rotating annulus I. A test of local stability analysis, Geophys. Astrophys. Fluid Dynam. 27, 137-162, 1983.

Fearn, D.R., Hydromagnetic waves in a differentially rotating annulus II. Resistive instabilities, Geophys. Astrophys. Fluid Dynam. 30, 227-239, 1984.

Fearn, D.R., Hydromagnetic waves in a differentially rotating annulus III. The effect of an axial field, Geophys. Astrophys. Fluid Dynam. 33, 185-197, 1985.

Fearn, D.R., Hydromagnetic waves in a differentially rotating annulus IV. Insulating boundaries, Geophys. Astrophys. Fluid Dynam., in press, 1988.

Fearn, D.R., Roberts, P.H. and Soward, A.M., Convection, stability and the dynamo, in the proceedings of the conference on Energy, Stability and Convection, Capri, 1986 (B. Strauhghan and P. Galdi, eds.), Longmans, Harlow, 60-324, 1988.

Jacobs, J.A., Reversals of the Earth's Magnetic Field, Adam Hilger, Bristol, 1984.

Michael, D.H., The stability of an incompressible electrically conducting fluid rotating about an axis when current flows parallel to the axis, Mathematika 1, 45-50, 1954.

Roberts, P.H. and Loper, D.E., On the diffusive instability of some simple steady magneto-hydrodynamic flows, J. Fluid Mech. 90, 641-668, 1979.

Soward, A.M., Thermal and magnetically driven convection in a rapidly rotating fluid layer, J. Fluid Mech. 90, 669-684, 1979.

Wesson, J.A., MHD stability theory, in Plasma Physics and Nuclear Fusion Research (R.D. Gill, ed.), Academic Press, London, 1981.